Reflections

The Magic, Music and
Mathematics of
Raymond Smullyan

Reflections

The Magic, Music and
Mathematics of
Raymond Smullyan

Raymond Smullyan

Indiana University, USA

World Scientific

NEW JERSEY · LONDON · SINGAPORE · BEIJING · SHANGHAI · HONG KONG · TAIPEI · CHENNAI

Published by

World Scientific Publishing Co. Pte. Ltd.

5 Toh Tuck Link, Singapore 596224

USA office: 27 Warren Street, Suite 401-402, Hackensack, NJ 07601

UK office: 57 Shelton Street, Covent Garden, London WC2H 9HE

Library of Congress Cataloging-in-Publication Data
Smullyan, Raymond M.
   Reflections : the magic, music, and mathematics of Raymond Smullyan / by Raymond Smullyan (retired).
     pages cm
   Includes bibliographical references.
   ISBN 978-9814644587 (hardcover : alk. paper) -- ISBN 978-9814663199 (pbk : alk. paper)
   1. Smullyan, Raymond M. 2. Mathematicians--United States--Biography. 3. Logicians--United States--Biography. 4. Magicians--United States--Biography. I. Title.
   QA29.S647S68 2015
   510.92--dc23
   [B]

                                 2015001067

British Library Cataloguing-in-Publication Data
A catalogue record for this book is available from the British Library.

Typeset by Stallion Press
Email: enquiries@stallionpress.com

Printed in Singapore

Foreword

At one university at which I gave a lecture, I opened by saying, "Before I begin speaking, there is something I would like to say." This, of course, got a general laugh. I then explained that I did not invent that lovely phrase, but that it was due to the late computer scientist Saul Gorn, who used it as a part of a book of self-defeating sentences, whose title was *Saul Gorn's Compendium of Rarely Used Clichés.*[1] The book contains such items as:

(1) These days, every Tom, Dick and Harry is named John.
(2) Half the lies they tell about me are true.
(3) I am a firm believer in optimism, because without optimism, what is there?
(4) This book fills a long needed gap.
(5) If Beethoven were alive today, he would turn over in his grave!
(6) The reason I don't believe in astrology is that I'm a Gemini!

I then introduced myself by means of what I would call a "meta-introduction" — that is, I told the group about three introductions I had had in the past. The first was by my former student, the now retired Professor Melvin Fitting. Before I tell you this introduction, I must give you some background: In one of my puzzle books, I give a proof that either Tweedle Dee exists or Tweedle Dum exists, but there is no way to tell which. In another of my books, I sort of "doubled up" on a famous result of the great logician

---

[1]This entertaining book can be downloaded at the following website: http://repository. upenn.edu/cis_reports/501/

Kurt Gödel, who showed that for even the most advanced mathematics axiom systems of our time, there must always be sentences of the system which are true but not provable in the system. Well, I constructed a very simple system in which I exhibited two sentences X and Y such that one of the two must be true, but not provable in the system; but there was no way to tell which of the two it was. Well, it was things like this that once induced Melvin to introduce me at a math lecture I was about to give by saying, "I now introduce Professor Smullyan, who will prove to you that either he doesn't exist or you don't exist, but you won't know which!"

In another introduction I had, the introducer at one point said, "Professor Smullyan is unique." Well, I was in a mischievous mood at the time, and so I interrupted him and said, "I'm sorry to interrupt you, sir, but I happen to be the only one in the entire universe that is not unique."

The third introduction was particularly delightful and didn't concern me in particular, but could be given for anyone. It was by the philosopher Nuel Belnap Jr. who said, "I promised myself three things in this introduction: One, to be brief; two, not to be facetious, and three, not to refer to this introduction." After this "meta-introduction", I then said to the group, "I have two different possible lectures prepared for you tonight, and I want you to decide which you would rather have me give. One of them is very impressive, and the other is understandable."

This really got a hearty laugh!

To depart from my main topic for a moment (which I will frequently do in the book!) I must tell you that my joke above contains a serious element: It is surprising how many people identify obscurity with profundity! I recall one occasion when a lecture was given at a university. After it was over, I heard one member of the audience say to another, "This lecture couldn't be very good. I understood everything!"

On another occasion, while I was teaching at Princeton, a very brilliant graduate student from another university (who later became a famous logician) was visiting Princeton and asked me whether I would read a paper he was planning to submit to a logic journal. I took the paper home and had a terrible time understanding it! After much trouble, I finally realized the essential idea, which was a beautiful one, but so clumsily expressed! In a fraction of the space, he could have given an elegant and easily understandable explanation of his brilliant idea. The next day, when I told this to him, he looked thoughtful and said, "No, no. If I did it that way, people would think that it was trivial!"

Another relevant incident: someone once sent in a review of a paper that I had written. The review editor wrote back to the reviewer that she had better delete the last sentence since the author of the paper (me) might find it insulting. Fortunately, the last sentence was not deleted, and far from regarding it as insulting, I regarded it as most complimentary! The sentence is, "This paper is easy to read."

After this long and typical ramble, I must tell you that after one lecture, a member of the audience came up and told me that he and several of his friends were fans of mine and would like to know more about my life and thoughts. He asked me, "Couldn't you write a memoir?" Well, there is already my book of 2002 entitled *Some Interesting Memories*, but many interesting things were left out, not to speak of the equally, if not even more, interesting events of my life from 2002 to the present (early 2015). Thus, I decided to do a more comprehensive job, hence the present book.

Contents

Contents

1 | Early Childhood

As I mentioned at the end of the Foreword, I decided to write this book after some readers of my previous books expressed the desire to know more about my life and thoughts. So here goes!

As this work goes to publication in 2015, I am currently 95 years old and reside in a beautiful region of the Upper Catskill Mountains — close to a small village named Tannersville. Tannersville is part of the town of Hunter, which has a famous ski slope. I live in a home built in 1966, which was designed by my late wife Blanche who passed away in January 2006 at the ripe old age of 100! We were married for 48 years.

As you might guess from the fact that I started this memoir by telling you a little about my life in 2015, I shall not be at all systematic in my account, but shall ramble along as my fancy dictates. That is, if in relating an incident in my life, I am caused to suddenly recall something much later, I shall interrupt my narrative and jump ahead to the later event, and this in turn might remind me of some other incident or some joke, or some riddle or puzzle, and so I shall pursue that before returning to the narrative of my life. Some readers might find this disturbing while others may find it refreshing. At any event, let me begin.

I was born in 1919 in Far Rockaway, New York, a place I sadly miss! My father was Isadore Smullyan, who I believe was born in Russia and immigrated to Belgium at an early age. His native language was French. My mother, Rosina Smullyan, neé Rosina Freeman, was born and raised in

London, and came to America shortly after her marriage. She was a painter whose paintings were all unfortunately lost. She was also an excellent Shakespearean actress (not professional), who gave memorable recitations on various occasions.

Both of my parents were musical — my father played the violin and my mother the piano. I was the youngest of three children. The eldest, ten years my senior, was Emile Benoit Smullyan, who later became an economist under the name of Emile Benoit. He authored two books, *The Common Market* and *Europe at Sixes and Sevens*. The middle child was Gladys Smullyan, later known as Gladys Gwynne, who was six years older than me.

My earliest recollection is of being in my crib and being disturbed by the color of the wallpaper!

A particularly significant (and somewhat humorous) incident occurred when I was two or three years old, which gave a good indication of what I would be like as an adult. I used to sit on my grandfather's lap and liked to play with the smoke coming from his cigarette. On one occasion when I was on his lap, he was not smoking and I wanted him to. And so I kept saying, "Moke! Moke!" He ignored me, and so I continued saying, "Moke! Moke!" After much of this, he decided to distract me by telling me a long, long story. I listened quietly and attentively to the story and when it was over, I said, "Moke! Moke!" He smiled, turned to my mother and said, "I see what he will be like when he grows up!" My friends, to whom I have told this, have said, "You haven't changed a bit!"

My mother told the family that a few months prior to my birth, she prayed that I would play the violin, like my father, and be born on his birthday. Well, I was indeed born on his birthday and did play the violin as well as the piano until the age of sixteen or so, when I gave up playing the violin and continued with the piano. My mother also told me that she had quickly recognized I had musical talent because when I was only a few months old and was taken outdoors in my crib, I would listen to the birds singing and would sing back the very same notes!

In addition to my mother and me, my brother and sister also played the piano. I would listen attentively when any of the older family members would play, and would later go to the piano and experiment with various keys. I would also improvise, and one of my early improvisational compositions I titled "Happiness". One day, while my mother was on the phone, within earshot and view of the piano, I went to the piano and by ear I played the first half of "My Country Tis of Thee". My mother smiled, and when she finished the phone call, she went over to the piano and said to me, "Let's play a little game. You turn your back and I will strike some notes on the piano, and see if you can tell me what they are." (I had already learned the names of the notes.) She did this and I named all of the notes perfectly! My mother thus discovered that I had perfect pitch (also called "absolute pitch"). As a child and a young man, I did indeed have an excellent sense of absolute pitch which I gradually lost during my later years. These days, I hear all the notes about three tones too high — for example, I hear C as an E-flat. I understand that the great pianist Alicia de Larrocha had the same problem. As I recall reading somewhere, she wrote something like, "If someone plays a note and asks me what it is, I reply that I hear it as an A and so I guess that it must be an F sharp."

When I was about eight or nine years old, we bought a new upright piano from a dealer in our town. He came to the house to tune the piano, but tuned it low, A435 instead of A440. When I tried playing it, I was terribly upset! I told the tuner that I wanted it at 440. He was certainly not knowledgeable, for he said, "Perhaps you want it at C440?" At the time, I did not know that 440 referred to A, not to C. I told my violin teacher at the time about this, and he said, "That is ridiculous: If C was 440, C would be A and A would be F#! Tell the man to either tune it right or to take the piano back!" I later tried to straighten it out with the dealer and at one point he said, "Perhaps you want it at concert pitch?" I wasn't certain at the time what concert pitch was (it was indeed A440!) and so I said, "Try that." He did and it sounded perfect! Thus, the incident got happily resolved.

To digress from my narrative a bit, I would like to tell you two stories about absolute pitch. Is it possible for someone who has had no musical training whatsoever, and who has never learned the names of the notes, to be known by others to have absolute pitch? The answer is YES! I knew a police officer who was totally unmusical, and never knew the names of the notes, who nevertheless was known to have absolute pitch. How? Well, this was eighty-five years ago, when police stations emitted radio signals, each station having its own individual frequency. The police officer in question was the only one among his fellow officers who upon hearing the radio signal could identify the police station!

Speaking of policemen, a certain police officer was related to a family that I frequently visited. On one occasion, when told of my interest in logic, he told me the following: "Here is my idea of logic. My wife and I once came late to a party. The hostess presented us with a plate on which there were two pieces of chocolate cake. One was larger than the other. I was closest to the plate and had to make the choice. Well, I reasoned as follows: I like chocolate cake, my wife likes chocolate cake, my wife loves me and wants to make me happy, and so I took the larger piece."

This brings to mind a few jokes: Two men went into a restaurant and ordered food. The waiter brought a platter on which were two pieces of fish — one larger than the other. One of the men said to the other, "Please, help yourself." The other said "OK," and took the larger piece. An angry silence followed. The other finally said, "You know, if you had told me to choose, I would have taken the smaller piece!" upon which the other said, "You have it, don't you? What are you complaining about?"

Then there is the story of a banquet at which a tray of asparagus was passed around. When the tray came to one woman, she cut off all the tips, put them on her plate, and passed the rest to her neighbor. The neighbor angrily said, "Why do you do a thing like that? Why do you keep all the tips for yourself and pass the rest on to me?" The woman replied, "Oh, the tips are the best part, didn't you know?"

The third joke is a cartoon I once saw in which a little boy and girl were walking on the sidewalk and the boy was walking inside. A truck then passed by and sprayed mud all over the girl's dress. The boy said, "Now do you understand why I don't walk on the outside like a gentleman?"

Coming back to the second story I want to tell you about absolute pitch, here is the funny incident in question. I am a close friend of the famous computer scientist Marvin Minsky, who is married to a second cousin of mine. One day I was in the front seat of a car that Marvin was driving. In the back seat were two men from Bell Labs. The conversation turned to absolute pitch.

Marvin said to the two in the back, "You know that Raymond has absolute pitch." One of the two asked me, "How accurate is your sense of absolute pitch?" For some odd reason, I didn't hear the question. He then repeated it louder, at which point Marvin, with his typical sense of humor, said to the two in the back, "Oh, I forgot to tell you that Raymond is also deaf!"

Marvin does indeed have quite a sense of humor, as the following incident will reveal. I published my first logic paper at the age of thirty-six. Someone who had read the paper said to Marvin, "That's a great paper that Raymond has written!" Upon hearing this Marvin said, "Oh yes. At the age of thirty-six, Ray decided to become a child prodigy!"

Speaking of logic, this brings me back to the mainstream narrative of my life. My introduction to logic was at the age of six. It happened this way: I was in bed with a cold, or flu, or grippe ... I don't remember what, and it was April 1st. In the morning, my brother Emile came to my bed and said, "Today is April Fool's Day, and today I will fool you like you have never been fooled before!" Well, I waited all day long for him to fool me, but he didn't. When night came, my mother came into my room and asked me why I hadn't gone to sleep. I replied, "I'm waiting for Emile to fool me." My mother had Emile come into my room and asked him, "Emile, why don't you fool the child?" Emile then asked me, "You expected me to fool you, didn't you?" I replied, "Yes!" He said, "But I didn't, did I?" I said, "No." He continued, "But you expected me to and I didn't, so I fooled you, didn't I?"

Well, after my mother and brother left the room, I recall being awake for quite a while, wondering whether I had really been fooled or not! On the one hand, if I wasn't fooled, then since I had expected to be, and my expectation was not realized, then I was fooled, which is a contradiction! On the other hand, if I was fooled, then I did get what I expected. So in what sense was I fooled? I was really most perplexed by all of this, until I finally fell asleep.

Many years later, when I was a mathematics professor, I told one of my colleagues about this childhood experience, and he said, "That might well explain your persistent interest in Gödel's theorem!"

Let me now tell you a little about this remarkable theorem. At about the turn of the 20th century, there appeared two mathematical systems that were so extensive that it was generally assumed that all mathematical questions could be answered affirmatively or negatively from the axioms of the systems. That is, given any mathematical statement, it could either be proved or disproved within the systems. But in the year 1931, the logician Kurt Gödel startled the entire mathematical world by showing that this was not the case. He showed that for the two systems in question, as well as for a variety of other mathematical systems, there must always be sentences which, though true, could NOT be proved in the system! Roughly speaking, the sentence he exhibited in his proof of this claim asserted its own non-provability in the system! To come closer to what Gödel said, he assigned to each sentence of the system a positive whole number, now known as the Gödel number of the sentence. He then constructed a most ingenious sentence G that asserted that a certain number *g* was the Gödel number of an unprovable sentence. But *g* was the Gödel number of the very sentence G! Thus G asserted that its own Gödel number was the Gödel number of an unprovable sentence, which is equivalent to G asserting that G is not provable in the system. Thus, if G is true, then, as it correctly says, it is unprovable; whereas if it is false, then contrary to what it says, it IS provable in the system! Thus, there are exactly two possibilities:

(1) G is true but not provable in the system.
(2) G is false, but provable in the system.

Now, it was evident from the axioms of the system that only true sentences could be proved. No false sentences could be. This rules out possibility #2, hence the conclusion is that Gödel's sentence G is true, but not provable in the system.

In the introduction to Gödel's 1931 paper, he said, "The paradox of the liar leaps to the mind." This paradox is about the Cretan who said: "All Cretans are liars." A better version is the following sentence: THIS SENTENCE IS FALSE.

Is the above sentence true or false? Suppose it is true. Then, as it correctly says, it is false, and we thus have a contradiction. On the other hand, suppose the sentence is false. Then, contrary to what it says, it is not false, which means that it is true, and we again have a contradiction. So, either way, we have a contradiction.

I hope you can see the connection between the above paradox and the paradox that arose when my brother Emile pulled his prank on me, which led me to believe that if I was fooled, then I wasn't; and if I wasn't, then I was! That certainly is a paradox, isn't it?

These paradoxes are related to an incident that occurred to me many years later: I was about thirty years old, and was earning a living as a magician. At one point, my business was very slow, so to augment my income, I applied for a job as a salesman. I had to take an examination and one of the questions was: "Do you object to telling a little lie now and then?" Well, I certainly did, and still do object to a salesman lying and misre-presenting his product, but I thought that if I truthfully answered "yes", I wouldn't get the job, and so I lied and answered "no". Many hours later, I wondered whether I objected to the lie I had told the sales com-pany. I came to the conclusion that I didn't object. Then I realized that since I didn't object to that particular lie, it then follows that I don't object to all lies, hence my answer "no" on the test was not a lie, but the truth! So was I then lying or not?

To this day, I do not know the answer to this paradox! All I can say is that at the time that I answered "no" to the sales company, I sure *felt* that I was lying!

Speaking of lying, there is the incident of the philosophers Bertrand Russell and G. E. Moore. Russell described Moore as one of the most hon-est people he had ever met. He once asked Moore whether he had ever lied. Moore replied, "Yes." In describing this incident, Russell wrote, "I think this is the only lie Moore ever told!"

I'm suddenly reminded of a story. A man asked an elderly gentleman, "You look so young and healthy for a man your age. What is your secret?" Whereupon the elderly gentleman replied, "It's because I never argue." The other man responded, "Oh, I bet you *sometimes* argue!" The elderly man replied, "Maybe you're right!"

I thought of the following *dual* version of the above story: One man says to a friend, "I think it's alright to argue sometimes." The friend replies, "I disagree!"

Coming back to the subject of lying, lying does not consist of making a false statement, but making a statement that one *believes* to be false. Indeed, if a person makes a *true* statement which he believes to be false, he or she is lying. This is nicely illustrated by the following true incident that I read about in an abnormal psychology book. The doctors of some mental institution were thinking of discharging a certain schizophrenic patient. But they decided to first give him a lie detector test. One of the questions they asked him was, "Are you Napoleon?" The patient answered, "No." The machine showed that he was lying!

I'm reminded of the joke about a very pompous man who once visited a mental institution and got into an argument with one of the patients. At one point, he very stiffly drew himself up and said, "Do you know who I am?" The patient replied, "No, but they can tell you at the information desk."

This reminds me of the story of a lady who visited a mental institution and got into a conversation with one of the patients, who said, "I know that many of the patients here have told you the same story, but I must insist that I have been framed! I don't expect you to necessarily believe me, but I know you are the wife of the governor of this state, and all I want is that, while not necessarily believing me, he have my case reviewed. That's a reasonable request, isn't it? Won't you please tell the governor that all I ask is that my case be reviewed?" She thought this perfectly reasonable and agreed. She turned around and started to leave the room, upon which the man gave her a terrific kick in the rear and said, "And don't forget to tell the governor!"

As we are on psychiatrist jokes, I like the one in which the receptionist of a psychiatrist went into his office and told him that there was a man in the waiting room who thought he was invisible. The psychiatrist said, "Tell him I can't see him now."

There is also the story of a little girl who refused to come out from under her bed. Even food couldn't tempt her. The parents finally decided to call a psychiatrist in on the case. He came to the house, went into the girl's bedroom and had a very long consultation with her. He came out of the room looking very grave. The mother anxiously asked him what the problem was. He replied, "This is very strange. She seems to have the idea that if she should ever venture out of her hiding place, people around her would start biting her!" The mother replied, "Oh, is that all? In that case we will stop biting her!"

Then there is the incident of a man who went to a psychoanalyst with the following complaint. "When I am on top of the bed, I'm afraid that there is someone under the bed, and so I go under the bed. But then I become afraid that someone is on top of the bed, and so I go back to the top of the bed. But then I worry that someone is under the bed, and so I go back and forth all night and don't get any sleep!" The psychoanalyst said, "Your problem is curable. It will take about two years. You will have to come and see me five days a week. I charge one hundred and fifty dollars per session." The man replied, "I'll have to think about it." Some months later, the psychoanalyst met the man on the street and asked him what he had decided to do. The man told him. "I took the problem to my bartender and he told me that he would give me a perfect solution for ten dollars. And he then gave me a solution which indeed worked perfectly. He told me to take the legs off the bed!"

Coming back to lying, I read somewhere the following incident showing how animals can sometimes dissimulate: An experiment was conducted on a chimpanzee in a room in which a banana was suspended by a string from the center of the room. The banana was too high for the chimp to reach. However, there were broken boxes of various sizes about the room. The purpose of the experiment was to determine whether or not the chimp was clever enough to make a scaffolding of the boxes, climb up and reach the banana. The experimenter stood in one of the corners to see what the chimp would do. Well, what happened was that the chimp came over to the corner and anxiously tugged the experimenter by the sleeve, indicating that he wanted the experimenter to move. Slowly, the experimenter followed the chimp. When they both reached the middle of the room, the chimp suddenly jumped on his shoulder and got the banana!

After this long detour, let me get back to the mainstream events of my own life. After having discovered that I had absolute pitch, my mother and father took me to the Juilliard School of Music in Manhattan for an interview with an assistant of the pianist Harold Bauer named Miss Pyle (a very charming lady I was later told). After my mother proudly told Miss Pyle that I had perfect pitch, Miss Pyle told her, "Most of the students here do." The upshot of the conversation was that Miss Pyle said, "Give him a fiddle!" Well, I didn't get a violin until a couple of years later.

I started piano lessons at the age of six with one Miss Daniels, who also taught my sister Gladys. I don't really know who taught my brother Emile, who sat improvising at the piano for hours at a time. Two years later, we moved to another house in Far Rockaway and I was taken to a music studio owned by two brothers, Victor Huttenlacher, who taught me the piano, and Ronald Huttenlacher, who taught me violin. I studied with them both for about five years.

Speaking of the violin and piano, I must ramble again with some jokes.

A man asked a friend, "Can you play the violin?" The friend replied, "I don't know, I never tried!"

A man went to a doctor for a broken arm. Sometime later, when the arm was healed, he asked the doctor, "Can I now play the violin?" The doctor replied, "I don't see why not." The man said, "Funny, I never could before!"

A man had an eight-year-old boy who played the piano, and the father was very fond of showing him off. After the boy had finished playing one evening, the father beamed with pride, and asked the company, "Well, what do you think of his execution?" One man replied, "I'm all for it!"

A man once took his dog to a theatrical agent. The dog took a violin out of its case and played. The agent shook his head, and said, "A Heifetz, he'll never be!"

That one is somewhat similar to the one about a man who visited a friend, and to his amazement, found his friend playing chess with his dog. When the man expressed his amazement, the friend said, "He's really not that good. I beat him two times out of three!"

The following incident is true: The great violinist and renowned teacher Joseph Gingold once played the following April Fool's joke on a student. Before the student came into the studio for a lesson, Gingold said to the accompanist at the piano, "Transpose the accompaniment a half tone low!" When I congratulated Gingold for this lovely prank, Gingold said: "The idea was not original with me. For years I was concert master of the Philadelphia Orchestra under Eugene Ormandy. Now Ormandy has absolute pitch. On one April Fool's Day, the orchestra decided to play a joke on him and so before he came into the rehearsal hall, they all tuned their instruments a half tone low!"

I am very fond of April Fool's jokes. One of my favorites is what might aptly be called a "meta" April Fool's joke. When I taught at Indiana University, the chairman of the philosophy department had two children — Johnny, aged 10, and Jennifer, aged 8. One April Fool's Day, Johnny came down in the morning and pulled one April Fool's joke after another on his parents. Then, little Jennifer came down and Johnny tried an April Fool's joke on her. She then said, "What's the matter with you, Johnny? Today's not April Fool's Day!" Johnny, amazed said, "It isn't?" Upon which, Jennifer said, "April Fool!"

Clever girl! This brings to mind some other clever children I have known.

When I was a graduate student at Princeton University, I had an office mate, Barry Mazur, later a famous mathematician at Massachusetts Institute of Technology. Some years later, after I had published my first book of logic puzzles, I got a letter from Barry's ten-year-old son Zeke. The boy proposed a wonderful logic puzzle that gave me an idea for a whole chapter of logic puzzles! I phoned Barry and wanted to talk to Zeke and congratulate him for his clever puzzle. Before Barry put his son on the phone, he said to me in soft, conspiratorial tones, "Look, Zeke is reading your book and loves it. But don't let him know it's math, because he hates math!" I find that most revealing!

Next, when I was a student at the University of Chicago, a fellow student had two brothers aged 8 and 6. I frequently went to their house and entertained them with magic tricks. On one occasion, I tried to scare them a little bit by saying: "I have a magic trick that would turn both of you into lions!" To my surprise, instead of being frightened, one of them said, "OK, turn us into lions!" I said, "Well, uh ... really, uh ... I shouldn't do that, because there is no way that I can turn you back again!" The little one said, "I don't care; I want you to turn us into lions anyway!" I replied, "No, really, there is no way I can turn you back." The older one shouted,

"I WANT YOU TO TURN US INTO LIONS!!" The little one then asked, "How can you turn us into lions?" I replied, "By saying the magic words." One of them asked me, "What are the magic words?" I replied, "If I said them, you would turn into lions." One of them then asked me, "Aren't there any magic words that would bring us back?" I replied, "Yes, but the trouble is this, if I said the magic words, then not only you two, but everyone in the world, including me, would turn into lions. And lions can't talk, and there would be no one left to say the other magic words!" The older one then said, "Write them down!" The little one said, "But, I can't read!" I said, "No, no! It wouldn't work. Even if the magic words were written down, everyone in the world would turn into lions." They responded, "Oh."

A few days later, I met the older one, who said, "Smullyan, there is something that I've been wanting to ask you — something that has been bothering me." I replied, "Yes?" He said, "How did you ever learn the magic words?"

He really had me stumped!

There was another occasion in which I was outwitted by a boy aged nine and a half, but I will tell you about this later.

12

2 | Later Childhood

Now, back to my own childhood. In my years 8–13, my main interest was science — particularly chemistry. My brother gave me a chemistry set and I augmented it with several more chemicals and converted one of the unused top floor bathrooms into a good laboratory. Some of the chemicals were given to me by a chemist, Bernard Wager, one of my brother's friends, who once told me of an incident that puzzles me to this day. He said that he once washed his hands in 100% concentrated sulfuric acid, and because there was no water present, it was perfectly safe to do so. Is this really possible? Is complete dehydration adequate to prevent ionization?

I was also fond of electronics in those years and I built my own radio from used parts that I purchased at radio stores. One radio store in particular gave me lots of parts without charging me. When I built the radio and showed the chassis to my father, he looked at the mess of wiring and said, "This won't work in a million years!" Well, I had no batteries to test it, but when I took it to a friend, a radio enthusiast who had the right batteries, the radio worked perfectly!

A very close friend of mine, Bernard Horowitz, lived next door, and was a real science buff. He later had many patents to his name. He and I stretched a long wire from a window at my house to a window of his, and we used to communicate with each other by Morse code.

The years between my eighth and thirteenth birthdays were probably the most interesting years of my childhood. I am very fortunate to have come from a cultured family. We always had dinner together, and I will never forget the interesting conversations that went on, both when we were alone and when we had relatives and guests for dinner. The conversations were usually of a somewhat philosophical nature. I recall one particular conversation, in which my uncle and aunt were present, and in which we spent hours discussing what was meant by saying that an individual has personality!

During those years, there was a 20-volume encyclopedia in the house entitled *The Book of Knowledge*, published by Grolier. I believe I learned more from reading those books than I learned from all my years in grade school. I don't know how the modern edition of those books is, but the edition I had at the time was superb, and one I most highly recommend! I believe my interest in logic puzzles stems largely from this very source. I recall the following nice puzzle. What happens if an irresistible cannonball hits an immovable post? By an irresistible cannonball is meant a cannonball that knocks over anything that it hits, and by an immovable post is meant a post which cannot be knocked over by anything! And so, what happens if this irresistible cannonball hits the immovable post? I'll give you the answer shortly.

Another puzzle I got from *The Book of Knowledge* is one that usually stirs a lot of controversy! It is about a man who is looking at a portrait. A friend nearby asked him whose portrait the man was looking at. The man replied, "Brothers and sisters have I none, but this man's father is my father's son." Whose portrait was the man looking at? Most people give the wrong answer. I'll give you the correct answer shortly.

Now to answer the problem of what happens if an irresistible cannonball hits an immovable post: The answer is that it is logically impossible for there to simultaneously exist an irresistible cannonball and an immovable post. If a cannonball is irresistible, then by definition, it will knock over any post it hits; hence no post can possibly be immovable!

This problem reminds me of the problem about the male barber of a certain town that shaved all of the inhabitants of the town who didn't shave themselves but never shaved any inhabitants who shaved themselves. Did the barber shave himself or didn't he? Well, suppose he shaved himself? Then he shaved someone who shaved himself — namely the barber. But this is contrary to the given condition that he never shaved anyone who shaved himself. Thus, it cannot be that the barber shaved himself. On the other hand, suppose he didn't shave himself. Then he failed to shave someone (namely himself) who didn't shave himself, so again this is contrary to the condition that he shaved all the inhabitants of the town who didn't shave themselves and so we again have a contradiction. How can this be? I'll give you the answer shortly.

Now for the problem of the man looking at the portrait who said, "Brothers and sisters have I none, but this man's father is my father's son." Whose portrait was he looking at? Most people give the incorrect answer that he was looking at a portrait of himself, and it is quite difficult to convince them that they are wrong. They reason that since the man has no brothers or sisters, then his father's son must be himself. So far, that is correct — the man's father is indeed himself, but that is not the answer to the problem! The man did not say "This man is my father's son"; he said, "This man's *father* is my father's son!" If he had said, "This man is my father's son" then it would be true that he was looking at a picture of himself, but to say "This man's *father* is my father's son" is equivalent to saying, "This man's father is myself", and thus the man is looking at a portrait of his son. If some of you are still not convinced, picture it this way:

THIS MAN'S FATHER IS <u>*MY FATHER'S SON*</u>
(*myself*)

In other words, in the man's statement simply substitute "myself" for the more complex phrase "my father's son". Now do you see it? If not, there is nothing more that I can say to help you.

Now let's get back to the barber problem. The solution is so simple that it gets overlooked! Look, suppose I told you that a certain man is more than six feet tall and also less than six feet tall. How would you explain that? The correct explanation is that if I told you that, I would either be mistaken or lying! There simply cannot be such a man. And so it is with the barber, there simply cannot be such a barber. In telling you that there was such a barber, I was giving you contradictory information.

I must tell you of a funny incident: I gave the barber paradox to a friend of mine — a pianist named Monica Alianello. With her typical sense of humor, she suggested the solution: "He probably went to his brother's house in another town and shaved himself!"

The following puzzle is, I think, a good introduction to logic. It concerns three people, Arthur, Betty and Charles. Arthur is looking at Betty and Betty is looking at Charles. Arthur is married and Charles is not. Does it follow that one of the three is married and looking at an unmarried one?

Interestingly enough, when the problem was given to me, I didn't solve it! I thought that since it was not given that Betty was married or not, then, not enough information was given to determine whether or not one of the three is married and looking at one who is not married. Also, one of my brightest graduate students, now a Ph.D., couldn't solve the problem. The solution is really so simple when looked at the right way. Either Betty is married or she isn't. If she is married, then Betty is looking at unmarried Charles. If she is not, then married Arthur is looking at unmarried Betty! Thus, in either case, a married person is looking at an unmarried one.

That bright former student of mine I just mentioned was amazed when I told her that I hadn't gotten the solution either! She said, "But the principle is one which you used over and over in the puzzles that you have published!" That is indeed true, but it just never occurred to me to split the problem into two cases, or as the author of the problem would put it, "to use *disjunctive* reasoning". The fact that I couldn't solve the problem worried me for a while, until I found out that some research was done, which concluded that there was practically no correlation between the ability to solve this problem and intelligence. Since learning this, I have given the problem to several others who were very bright but couldn't solve this problem, and others, much less bright, who quickly solved it. What is the characteristic that determines whether a person solves the problem or not? If it is not intelligence, then what is it?

Here is another little logic problem to introduce people to the subject of logic. I once came across a sign in a restaurant which said:

"GOOD FOOD IS NOT CHEAP.
CHEAP FOOD IS NOT GOOD."

Do those two sentences say the same thing or different things? Many people to whom I have posed this have said that they say different things. Well, *logically* they say the same thing — namely that no food is both good and cheap, but *psychologically,* they evoke quite different thoughts: When one reads "Good food is not cheap", one tends to think of good, expensive food, whereas "Cheap food is not good" tends to make one think of cheap, rotten food.

Coming back to my childhood, even as a youngster, I was extremely fond of girls, and loved to show off before them. Indeed, I liked, and still like, to show off in general. I have frequently been criticized for being a "show off". Well, my response to that is that I believe that showing off is neither good nor bad in itself, but depends entirely on the quality of what the show off shows. If it is poor, then showing it off can be a bloody bore, but if the quality is good, then I say, showing it off is all to the good.

Well, as a child I loved to climb trees. Once, in the presence of several girls, I climbed a tree, and while climbing said, "I'm the best tree climber around here!" Hardly had I finished saying that, when I fell to the ground! Fortunately, I was not hurt, but this is a delightful and quite literal case of pride coming before a fall! In addition to being criticized for being a show off, I have also been criticized for being immodest. Curiously enough, I have also been praised by others for being modest! In my honest opinion, I am quite immodest, but I'm not sure whether that is a bad thing! I love what Conan Doyle said about modesty through the mouth of Sherlock Holmes. In the story "The Greek Interpreter", at one point Watson says, "Holmes, I think you're being modest!" Holmes replies, "No, no Watson! I never regarded modesty as one of the virtues. To underrate oneself is just as much a departure from the truth as to overrate oneself!"

I once wrote the following dialogue on modesty:

A: For a person of your abilities and accomplishments you are remarkably modest!

B: I'm not modest!

A: Ah, I've caught you! By disclaiming your modesty, you are trying to show that you are so modest, that you won't take credit for anything — not even your modesty, but this is most immodest of you!

B: It's like I said; I'm not modest!

The following story is unfortunately not appreciated by everyone: A certain man was known as the world's most modest man. He signed all of his letters, "He who is modest." A theology student was once discussing this with his instructor and said, "How can he be modest when the very way he signs his name belies this fact?" The instructor said, "You don't understand! He is indeed modest! It's just that ever since modesty entered his soul, he no longer regards it as a virtue!"

Now, back to my childhood: When I was twelve years old, I entered a New-York-City-wide piano competition. The winners of the first round were awarded bronze medals and were allowed to participate in the second round — the semi-finals. The winners of those received silver medals and were allowed to participate in the finals. There was only one winner of the finals and he or she got a gold medal. I got a silver medal, but failed to get a gold medal. The next year I tried a second time, and before the day of the finals, a funny thing happened: For several weeks before that, my friend Bernard and I together built a canoe. Well, on the day before the finals, the canoe had been completed. I got up early in the morning, and while the rest of the family was asleep, went over to Bernie's house and the two of us carried the canoe onto the beach and into the water. We did not realize at the time that we were supposed to put lead on the keel to balance the canoe, and so the canoe kept turning over, dumping us several times into the water. When my parents heard about this, they were furious that I would do such a foolish thing the very day before a contest. Well, the next day, I entered the contest and won the gold medal!

The following year, when I was thirteen, my father lost all his money! He had had a thriving hardware business in Manhattan. The city had built a tunnel (perhaps related to the subway) and a lot of water filled the basement where all my father's hardware was stored and ruined it completely! My father sued the city, and amazingly enough, thirty witnesses for the city swore falsely, and my father lost the case! We had to sell our lovely home in Far Rockaway, and moved into the city to a small dismal flat. This began my high school years.

3 | High School

I could have gone to a very good high school fairly close to where I lived, but chose not to, because it wasn't co-ed. Instead, I went to a distant one which was co-ed and offered a special music course. Curiously enough, the one subject I failed, both in high school and college, was, of all things, music! In high school, during a music appreciation course, the teacher announced that one of the girl students was about to sing, and asked if any student present could accompany her on the piano. I raised my hand and said, "I can. At least I think I can!" I went to the piano and accompanied the singer, to everyone's satisfaction. The teacher praised me highly and asked me then if I would play a solo, which I did. The teacher then invited me to her home, which was very close to where I then lived, and I visited her and her husband many times during the semester, where we played four hands together on the piano. Yet at the end of the semester, she failed me for reasons I still don't understand! Did I perhaps miss the final examination?

At that time I also played violin in the orchestra. At the end of the semester I failed that too! When I asked the teacher who was conducting the orchestra why my playing was not good enough, he surprised me by saying "You play very well!" I still can't figure this out!

While on the topic of strange occurrences in music courses I have taken, and jumping ahead in time for a moment: One semester when I was a student at the University of Wisconsin, I was taking a course in orchestration with Professor B., and was chosen to be the soloist with the university orchestra in the Beethoven Piano Concerto #1. I had many rehearsals, and the conductor, also Professor B., spoke very highly of my playing to several faculty members. The concert was a great success and I got rave reviews in the local newspaper. Nevertheless, at the end of the semester I failed the course! I also failed a course I took there in music appreciation!

Back in my high school days, I very much disliked my first year algebra course, which is rather surprising, since modern higher algebra is my favorite branch of mathematics, next to mathematical logic. My real love of mathematics started when I took geometry. The textbook used in the course was by Joseph P. McCormack, and was an excellent one, unlike the horrible textbooks used today. It followed Euclid's *Elements* quite closely. For the first time in my life, I understood what was meant by a logical proof! Many of the high school texts used today combine algebra and geometry in a single course, and give the student not the remotest idea of what is meant by a proof! But trying to fight the textbook lobby is about as hopeless as fighting the gun lobby! I really believe that the main cause of the poor education of students today is not the poor quality of the teachers, but of the textbooks!

Now I must tell you something curious. Towards the beginning of the geometry course, we were to prove that the base angles of an isosceles triangle are equal. Instead of the usual proof, which involves making a construction line, it occurred to me that the triangle would fit on itself, when flipped over in space — that is, let ACB be the triangle, with C the vertex and AB the base. Well, side AC is the same length as BC, hence would fit on BC, and BC would fit on AC, and the angle between sides AC and BC is the same as the angle between sides BC and AC, hence triangle ACB would fit on the triangle BCA! And so the two base angles must also be equal. The curious thing is that in an experiment in artificial intelligence, when a computer was asked to prove that the base angles of an isosceles triangle were equal, it came up with the very same proof I had!

Speaking of computers, do you know the story of the military computer? Well, when the Army first sent a rocket to the moon, a Colonel programmed two questions into the computer:

(1) Will the rocket reach the moon?
(2) Will it return to earth?

The computer thought for a while, and answered "yes". The Colonel was furious, because he did not know whether "yes" was the answer to the first question or to the conjunction of the two, and so he angrily wrote back, "Yes, what?" After a while, the computer answered, "Yes, Sir!"

Then there is the story of the salesman trying to sell a computer which he claimed knew everything. He asked a customer to ask the computer a question. The customer asked: "Where is my father?" The computer then answered: "Your father is now fishing in Canada." The customer said, "The computer is obviously no good! My father has been dead for several years." The salesman responded, "No, no! The question must be asked in more precise language. Let me ask the question for you." The salesman then asked the computer, "This man here, where is his mother's husband?" The computer replied: "His mother's husband has been dead for several years. His father is now fishing in Canada."

There is the story of the computer that was an expert in playing the stock market. When asked how to make money on stocks, the computer replied, "Buy low, sell high!"

Then there is the story of the computer who was asked whether there is a God. After a year, the computer said, "There is now!"

The following is true: An experiment was once performed in which an English or American idiom was translated into Russian by a computer, and then another computer translated the Russian version back into English to see how much distortion resulted. One of the idioms they tried was "The spirit is strong but the flesh is weak". What came back was "The vodka is good, but the meat is rotten". Another idiom they tried was "Out of sight, out of mind". What came back was "blind idiot"!

Let me now tell you some of my favorite math jokes:

A group of people got lost during a journey. They decided to test the echo from a nearby mountain and shouted, "We are lost!" Soon they heard back, "You are lost!" One of them said, "This was obviously not an echo, but a person." Another said, "Yes, a mathematician." When asked how he knew it was a mathematician, he replied, "For three reasons: First, he took a long time to answer. Second, his answer was correct. Third, it was totally useless!"

Perhaps my favorite math joke is about a physicist who came to the office of a friend of his, who was a mathematician, and told him, "I just made an experiment that proves conclusively that quantity **A** is bigger than quantity **B**." The mathematician replied, "That's perfectly understandable! You didn't even have to make the experiment. **A** must be bigger than **B** for the following reasons ... ." He then started to explain. At one point, the physicist interrupted him and said: "Just a minute; I made a mistake. It's **B** that's bigger than **A**!" The mathematician responded: "That's even more understandable! Here is why ... ."

Mathematics professors are reputed to be absent-minded. There is the incident of a mathematics professor who was walking down a hall and came across a student who asked him if he had already had lunch. The professor replied, "In which direction was I walking when you asked me the question?"

Another incident concerns a mathematics professor who in the course of a lecture, at one point, after stating a proposition, said, "This is obvious!" A student raised his hand and asked, "Why is it obvious?" The professor stood in thought for a couple of minutes, then left the room and paced up and down the hall for about twenty minutes, returned to the classroom and said, "Yes, it is obvious," and went on with the lecture.

The funniest absent-minded incident I know is about the celebrated mathematician Norbert Wiener. Whether or not the story is true, I cannot say, but it is quite plausible since Norbert Wiener had very bad eyesight in his later years. The story goes that the Wieners had to move from one end of Cambridge to another. Thirty days before the moving date, Norbert's wife decided to condition him in advance, and said, "In thirty days we will have moved, and so when you get out of class, you don't take Bus A; you take Bus B." Norbert replied, "Yes, dear." The next day, Mrs. Wiener said to Norbert, "Now in twenty-nine days we will have moved, and so when you get out of class, you take Bus B, not Bus A." Norbert again replied, "Yes, dear." This went on day after day, until the day on which Mrs. Wiener said, "Now Norbert, today is moving day, so when you get out of class, you take Bus B, not Bus A!" Norbert replied, "Yes, dear." Well, when Weiner got out of class, he of course took the wrong bus, returned to his former home, and found it empty! He then recalled that this was the moving day and so he thumbed his way back to Harvard Square, got on the right bus, got off at the right stop, but forgot his new address! He wandered around, and it was getting dark. He finally spied a little girl, went over to her and said, "Excuse me, but do you by any chance happen to know where the Wieners live?" She replied, "Oh, come on, Daddy. I'll take you home!"

* * *

Back to geometry: You recall the Pythagorean Theorem that in a right triangle, the square on the side of the hypotenuse is equal to the sum of the squares of the other two sides. I have a novel way of teaching this, which has proved very effective. I draw a right triangle on the board and draw squares on the hypotenuse and the two sides and I tell the class, "If these squares were made out of valuable gold leaf, and if you had the option of taking either the one big square or both of the small ones, which would you take?" Usually, about half of the class opts for the large one — the square on the hypotenuse, and both groups are equally surprised when I show them that it makes no difference!

I then tell the class the story of the American Indian Chief who had three squaws — one slept on the hide of a bear, one on the hide of a deer, and the third on the hide of a hippopotamus — sort of status symbols. The squaw on the hide of the bear had a daughter; the squaw on the hide of the deer had a son, and the squaw on the hide of the hippopotamus had twins, which proves that the squaw on the hide of the hippopotamus is equal to the sum of the squaws on the other two hides.

I used these pedagogical devices, not on high school students, but on first year remedial college students — students who never had an adequate mathematical background. I really learned so much in teaching the remedial students! It is incredible how many things that we take for granted are completely unknown to these students! For example, and this is hard to believe — it was a complete surprise to many of the students in one of my classes when I told them that if you divide a quantity into two equal parts, each part is half of the whole! On another occasion, as I gave a problem to the class, I said that there was a piece of string one hundred inches long. You cut off seven inches. How many inches are left? They couldn't get it! When I told them that the answer was ninety-three, several of the students said, "Oh, you get it by subtraction!" Now, had I asked them how much is one hundred minus seven, they would have gotten it, but they seemed to have no idea of how subtraction applies to the real world!

I'm a great believer in the value of common sense. On one final exam that I gave to a remedial class, I had one typical algebra problem involving finding the ages of a mother, father and child. I said to the class, "On this problem, I'll give you a hint!" All eyes turned to me eagerly. I said, "If the child turns out to be older than either parent, then you've done something wrong!"

Before coming back to my own high school days, I must tell you an amusing incident. A friend of mine told me that when he was in high school, he was doing very poorly in mathematics. As a result, he was sent to the principal's office, and in a thunderous voice, the principal said: "WHY ARE YOU DOING BADLY IN MATHEMATICS?" The boy replied, "Oh sir, I don't like mathematics!" The principal replied, "Oh, but you've *got* to like mathematics! If you don't know your mathematics, suppose you go into a grocery store and your bill is eighty-seven cents. You hand the clerk a dollar bill, and he gives you only thirteen cents — you wouldn't even know the difference!"

* * *

Now, back again to my high school days. I must tell you of one amusing incident in which I was quite unintentionally complimented. At the time, I was quite a fan of the philosopher and logician Bertrand Russell. Well, I had an English teacher who was a highly cultured lady but extremely conventional. We had to write essays, and I turned in one in which I expressed some very radical ideas. She returned my paper with the comment: "Your English is good but your thinking is confused! Please see me in my office." Well, the next day I went to her office and we had a long chat about various authors. At one point I asked her what she thought of Bertrand Russell. She angrily replied: "He's like you! His thinking is confused!" Boy did I feel good!

Incidentally, interestingly enough, I was once reading Bertrand Russell to my mother. At one point she said, "He reminds me of you!"

I must tell you that on another occasion, my mother wanted me to do something which I didn't want to do. When I told her that I wouldn't do it, and she told me that I was being selfish, I said, "Mother, for whose sake do you want me to be unselfish?" All my mother could say was, "You should really be a scientist!"

Coming back to my English class, we did a lot of Shakespeare, and here is a little ode to Shakespeare I just thought of, and on a more humorous note, it will be followed by a Shakespeare quiz I thought of many years later:

ODE TO SHAKESPEARE

Friends, Romantics, and Countrymen,
Lend me your Ears!
I come, not to bury Shakespeare, but to praise him.
The false words made by bards live after them,
The true are often interred with their bones.
And so it be with Shakespeare.
Many noble ones hath told you that William is exalted beyond
    compare!
'Tis a truth devoutly to be revered!
Shakespeare, by any other name, would be as great.
We are such stuff
As dreams are made of, and our little life
Is rounded with a sleep.
If poetry be the food of love, play on!

To be or not to be
A fan of Shakespeare.
That is the question!
Whether 'tis nobler in the mind to suffer
The slings and arrows of misfortune,
Or to take arms against a sea of troubles
And by opposing end them.
To die — to sleep —

To dream of William Shakespeare.
'Tis a consummation devoutly to be wished.
But then to wake, and be with him no more!
Ay, there's the rub!

Tomorrow, and tomorrow, and tomorrow.
Grows Shakespeare greater from day to day!
To the last syllable of recorded time,
The way to the man's mind.
Out, out, oh Chaucer!
Whom Shakespeare so dearly loved!
Life's but a poetic tribute
That has its hour upon the stage
And then is heard no more. 'Tis a tale
Told by a great sage,
Full of sound and beauty,
Signifying all that is wonderful!

A Shakespearean Quiz

(1) What Shakespeare play is about an ancient Roman general and ruler who had a bad cold?
(2) What Shakespeare play is about two cats?
(3) What Shakespeare play is about a lascivious king?
(4) What Shakespeare play is about a nocturnal howl?
(5) What Shakespeare play is about a romp of little fairy-like people after dark?
(6) What Shakespeare character is related to a tasty breakfast?
(7) What Shakespeare comedy is about tunes?
(8) What Shakespeare comedy is about something terrible?
(9) What Shakespeare historic play is about an eighth chicken jewel?
(10) What Shakespeare play is about a salesman of trouble?
(11) What Shakespeare play is about the sale of a beautiful goddess?
(12) What Shakespeare play is about the rehabilitation of wild footwear?
(13) What Shakespeare play is about the rehabilitation of a wild couple?
(14) What Shakespeare play is about the coloring of a wild rodent?
(15) What Shakespeare play is bound to us?

(16) What Shakespeare play is about a robot?

(17) What Shakespeare historic play concerns the energizing of a third battery?

(18) What Shakespeare play is about ten annoying insects?

(19) What Shakespeare historic play is about the beginning of a computer?

(20) What Shakespeare play is about a Greek salesman of neckwear?

(21) What Shakespeare play is the story of one who has won?

(22) What Shakespeare play is about two wedge-shaped blocks?

(23) What Shakespeare play is related to two Greek locks?

(24) What Shakespeare play is about a wicked Celtic king?

(25) What Shakespeare play delivers a resounding greeting?

(26) Which character in Macbeth sneezed a lot?

Answers

(1) Julius Sneezer

(2) Ro-Meow and Juli-Cat

(3) King Lear

(4) A Midsummer Night's Scream

(5) Elf Night

(6) Amlet

(7) Comedy of Airs

(8) Comedy of Terrors

(9) Hen-ring the 8th

(10) The Merchant of Menace

(11) The Merchant of Venus

(12) Taming of the Shoe

(13) Taming of the Two

(14) Tanning of the Shrew

(15) Tie-to-us Andronicus

(16) Titus Androidicus

(17) Re-charge the III

(18) The Ten-pests

(19) Mac-Birth

(20) Tie-man of Athens

(21) Winners' Tale

(22) Pair o'Cleats
(23) Pair o'Keys
(24) Sin-beline
(25) O-Hello
(26) MacSnuff

* * *

Coming back to my high school math classes, the third semester of algebra interested me much more than the first two semesters, largely because we then studied quadratic equations, which also lead to imaginary and complex numbers. I was then curious about how to solve the cubic (third degree) equation by radicals, and this was shown to me outside of school by a mathematician friend. I was also informed that the solution to the cubic equation was published by the Renaissance mathematician, Jérôme Cardan (or Gerolamo/Girolamo/Geronimo Cardano), who had a colorful life — a colorful life indeed! He was a medical doctor as well as a mathematician, an incurable gambler and possibly a murderer. He was tried for murdering his wife, but was acquitted. He did, however, cut off the ear of one of his sons, in one of his typical fits of rage! Now, this was the age of mathematical duels, and Cardan desperately needed to know the solution of the cubic equation to win his mathematical duels. The solution to this equation was discovered by Tartaglia, and Cardan kept begging Tartaglia to give him the solution, but Tartaglia kept it a secret. One day Cardan visited Tartaglia, fell on his knees and pleaded piteously. He swore that if Tartaglia gave him the solution, he would never, ever reveal it to anyone else! Tartaglia relented and gave him the solution. The next day, Cardan published the solution as his own!

The manner of Cardan's death was particularly curious! Towards the end of his life, he got interested in astrology and cast his own horoscope, which predicted the day that he would die. When that day came, Cardan committed suicide in order that the prediction be fulfilled!

My mathematician friend also showed me the solution of the fourth degree equation, which was discovered by a pupil of Cardan named Ferrari. "What about the fifth degree equations?" I asked. I was then told that no equation of the fifth degree or higher could be solved by radicals.

I was extremely curious to know why, and was told that I could find out by studying the subject known as *Galois Theory*.

The 19th century mathematician Évariste Galois was one of the most remarkable geniuses who ever lived! He was the virtual founder of the branch of mathematics known as *Group Theory*. He associated with every polynomial equation a certain group of substitutions of the roots — this group is now known as the *Galois Group* of the equation — and he showed that a polynomial equation has a solution by radicals if and only if its associated Galois group has a certain property known as *solvability*. He then showed that for the general equation of degree five or more, its Galois group is not solvable, and hence the equations of these higher degrees are not in general solvable by radicals.

Galois has been my mathematical hero and role model for many years! I have been as intrigued by his tragic life as by his remarkable accomplishments. All through his school and university days he was regarded by his professors as lacking in intelligence. The mathematical papers he submitted for publication were far beyond the comprehension of the judges who rejected them.

To digress for a moment, it is fascinating how mediocrity often reacts to genius! I understand that in the time of Johann Sebastian Bach, a certain church was looking for an organist. The minister of the church wrote to another minister, "If we can't get anyone better, I guess we'll have to settle for Bach." Also, when Beethoven once gave a concert in which he played his own compositions, one reviewer wrote, "He will never amount to anything as a composer!"

Galois was also quite politically active, he being an ardent republican as opposed to a monarchist. On one occasion he was arrested and tried for brandishing a knife and saying, "To Louis Philippe!" He was not convicted, since his defense attorney claimed that Galois had said, "To Louis Philippe, if he betray!" and that the last words were not heard because of the general noise of the occasion. However, he was arrested several months later and spent several months in jail.

At the young age of twenty, he was challenged to a duel by two monarchists. He spent the last night of his life, knowing that he would be killed the next day, feverishly writing his greatest work! The next day, he was found wounded and dying by his brother and his friend Chevalier, who then copied his mathematics papers, which were received by the mathematician Louisville, who published them in his mathematical journal in 1846.

It was partially my interest in Galois Theory and other branches of mathematics that caused me to drop out of high school in my junior year and to study on my own. I never did graduate from high school, but got into college by taking College Board Exams. Before entering college, I briefly attended evening classes at another high school, but soon dropped out of there too. I also dropped out of college several times. I guess that I am simply a born-again dropout!

During the years I was neither going to school nor working, I was generally regarded as a sort of idler and good for nothing! I recall that at one party that I attended in those days, someone asked me what I was doing with myself. I replied, "I'm waiting for the meek to inherit the earth!" At another party, a girl told me, "I don't believe your life is as empty as people think." She sure was right. Far from being idle ones, those days turned out to be at least as productive as any period in my life. In addition to learning Galois Theory, I learned analytic geometry and calculus completely on my own. I also then got interested in mathematical logic and independently discovered the subject known as *Boolean Rings*. I knew nothing about publication at the time, which was a good thing; if I had submitted my work on Boolean Rings for publication, it would have been rejected, since the subject had already been discovered some years earlier. Also, in those days, I composed numerous chess problems of a curious sort, which I didn't publish till many, many years later, when I was finally able to get them out in two books — *The Chess Mysteries of Sherlock Holmes*, and *The Chess Mysteries of the Arabian Knights*. How those two books got published is a remarkable story in itself, which I will tell you all about later on.

As I said, I got into college by taking the College Board examinations. I did extremely well in physics and chemistry, but, surprisingly enough, extremely poorly in mathematics! I simply blanked out during the algebra and geometry tests. I just sat there unable to think or write. Why this was so, I cannot explain!

30

4 | Beginning College

The first college I attended was a small one called Pacific University. It was located in Forest Grove, Oregon. Having studied freshman calculus on my own, I started in the second calculus course. I stayed at Pacific University for one semester and then transferred over to Reed College in Portland, Oregon, where I really should have started. The students at Reed College were far less provincial than those at Pacific University, and came from various parts of the country, as far away as New York (I was not the only one from New York). I remained at Reed for less than a semester as the result of an unexpected occurrence. Soon after I arrived Reed was visited by a well-known pianist from San Francisco named Bernhard Abramowitch. We struck up a close friendship, and he offered me a scholarship to come anytime to San Francisco to study with him. I could have waited until the end of the semester, but I was not too pleased with my studies, so I dropped out and went to San Francisco. [Yes, as I said, I really *am* a born-again dropout.] The year I spent in San Francisco and Berkeley was a particularly interesting one.

During the first half of the year, I spent many of my days at the house where Abramowitch was living. I so vividly remember waking up in the morning and hearing Bernhard practice those wonderful Schubert Sonatas. Schubert became my favorite Romantic composer. My two favorite composers — at least those I like to play best — are Bach and Schubert.

In the second half of the year I moved over to Berkeley, since I wanted to be near the University to audit courses. One particularly memorable incident concerns a course I audited in mathematical logic. It was a course in which we studied the *Principia Mathematica* of Whitehead and Russell. During the early part of the course, the students were having difficulty in finding proofs of various sentences of *propositional logic* — finding proofs, that is, from the given axioms. Now, there is a very simple mechanical method of determining whether or not these sentences are *valid* — this method being known as the *truth table method*. But verifying a sentence by a truth table is very different from finding a proof of the sentence from the axioms! Well, it occurred to me that there should be a mechanical method of converting a truth table for a sentence to a proof from the axioms of *Principia Mathematica*, and after some thought, I discovered such a method. I wrote it up, and the next day I handed it to the professor of the class. He handed it back to me, with the simple comment, "Correct!" I asked him whether he didn't think my result was publishable. He replied, "I don't think it would interest most logicians." Boy, was he wrong! He was right in that it was not publishable, but for very different reasons than those he had in mind. It certainly was, and is, of considerable interest to logicians, but it was indeed not publishable because it had already been published some years earlier by Emil Post. (This was the second time I had discovered something that was previously discovered by someone else.)

During my days in Berkeley, I met and became close friends with Leon Kirchner, who subsequently became a well-known composer. We had much fun together; in fact we took part in a play — a comedy titled "Petticoat Fever". We also gave a joint recital in Berkeley.

In my book *Some Interesting Memories* (Thinker's Press, Inc. 2002), various friends of mine contributed recollections of their relationships with me. Leon Kirchner wrote the following:

Raymond and I roomed together at Berkeley, California in 1939, when I was attending the University. We once gave a joint recital there. In 1942, Raymond and I were living in New York. At that time he was undecided whether to make music or mathematics his profession. I then knew the famous pianist Miezyslaw Horszowski and arranged for Raymond to have an audition with him. Raymond came back from the audition looking disappointed and said that Horszowski had said something to the effect that Raymond came a week too late. Well, many, many years later, when Horszowski was in his late nineties, I was having breakfast with him at the Marlboro Music Festival and somehow Raymond's name came up and I asked Horszowski whether he remembered him. He replied, "Oh yes, whatever happened to him?" I reminded him of the audition and of Raymond's disappointment at what he had said. He replied, "Oh no, he misunderstood me! What I had meant was that the registration at the Curtis School of Music had occurred a week before Raymond came to me, and had he come a week earlier, I could have arranged for him to get a scholarship there to study with Rudolph Serkin." I told this to Raymond a few weeks ago, wondering whether he would feel pleased or disappointed. He was actually very pleased.

In 1957, my wife and I once visited a colleague at the Institute for Advanced Study in Princeton, N.J. At one point, he told us he had to leave to go to a lecture. "In psychology?" I asked. "Oh no," he replied. "A group of us are going to a class given by a famous mathematician who is also a fabulous magician!" When I asked the name of the admired man, he replied, "Raymond Smullyan."

Now, while memories of my old friend Leon Kirchner are running through my mind, I want to mention something that occurred only much later, indeed after Leon had passed away in 2011. His biography, *Leon Kirchner: Composer, Performer, and Teacher*, by Robert Riggs was published in 2010 by the University of Rochester Press. A copy was sent to me by Leon's son Paul Kirchner, with the following lovely inscription:

May 12, 2011

To Raymond Smullyan,

Whom my father cherished as a lifelong friend and fellow musician and pianist and whose intellectual brilliance as an eminent mathematical logician my father greatly admired.

With warmest regards,

Paul Kirchner is primarily a painter, but also has a keen interest in mathematics. He has a particular interest in reviving old textbooks of exceptional quality (a most admirable project, considering the horrible math textbooks in use today!) He has edited and recommended several of them to Dover Press, which, as a result, reissued many of them. He also called Dover's attention to my books and is responsible for my present affiliation with Dover, which has proven to be most fruitful.

* * *

After my year in San Francisco and Berkeley, I returned to New York. Leon Kirchner soon came after me, and we were reunited. I'll tell you one memorable incident. Leon and I were particularly fond of the pianist Artur Schnabel. Once, Leon and I were listening to his magnificent recording of Schubert's great posthumous A-major sonata. We were both moved to the core. I jokingly suggested that we phone Schnabel and congratulate him. To my amazement, Leon picked up the phone and started calling! I rushed to an extension. After congratulating him on his performance, we told him that we had recently been to a concert of another famous pianist who had played the same sonata, but not nearly as well. Of course we were nervous about taking the time of the great Schnabel, but it was *he* who kept *us* on the phone for a long, long time, tracing the entire development of the sonata form!

I must tell you some more incidents about Schnabel. I visited him on two occasions, during one of which I played for him. He was in a very good mood and we had a long chat. At one point he said, "I am a realist! It is for that reason that I can sit back and be an idealist." When he saw me looking puzzled, he added, "Because ideals are the reality."

I attended some lectures that Schnabel gave at the University of Chicago. At one point, one of the members of the audience asked him what he thought of his latest review. Schnabel replied, "I don't read my American reviewers, because when they make a suggestion, I don't know what to do about it! Now, in Germany, it was different. On one concert that I gave, the reviewer wrote that I played the first movement of the Brahms sonata too fast. I thought about the matter and realized that the man was right. Well, I knew what to do — the next time I played it in public, I simply played it a bit slower. But when those American reviewers say things such as 'The trouble with Schnabel is that he does not put enough moonshine in his playing,' I simply don't know what to do about it!"

At another point of the lectures, Schnabel said, "It may be hard to believe, but Stravinsky actually published in the newspapers that music, to be great, must be cold and unemotional! And last Sunday I was having breakfast with Schoenberg, and I said, 'Can you imagine that Stravinsky actually said that music, to be great, must be cold and unemotional?' Schoenberg got furious and said, 'I said that first!'"

* * *

I remained in New York for a few years and then some friends of mine persuaded me to go back to school — specifically, to the University of Wisconsin, which they had attended and highly recommended. I did go to the University of Wisconsin. Shortly after I arrived, I phoned the eminent resident pianist Gunner Johansen and made an appointment to meet him in his studio at the university. I was hoping he could recommend me for an out-of-state scholarship. We talked in his studio for quite some time, and at one point he asked me on what grounds he could recommend me. I told him that I should play for him. He agreed, and after I had played for him, he said, "You play beautifully! I will certainly give you my warmest recommendation for a scholarship," which he subsequently did. Unfortunately, his recommendation, as well as the recommendation of two of the mathematics professors, failed to get me a scholarship! As the chairman of the scholarship committee told a psychology professor, "We wanted someone who would represent the spirit of Wisconsin!" Here is a related incident which I find quite revealing and rather funny: I had to interview with various members of the scholarship committee. In one case I was asked why I dropped out of high school. I replied that I got interested in a branch of modern higher algebra that was not taught in high school, nor in undergraduate college, and which I wanted to study on my own. The interviewer asked me, "Why didn't you follow the norms?" Really now! If a person's interests are not normal, which mine certainly were not (I am not saying that they were in any way *superior*; only that they were not normal), why should such a person follow the norms?

Coming back to Johansen, he was really quite upset when he heard that my scholarship was denied. As he told the professor of the psychology course that I was then taking: "When that boy sits down to play, it is sheer beauty!" What better compliment could I have had?

Now let me tell you something remarkable! At one point, Johansen told me, "If the university doesn't give you a scholarship this semester, I'll pay for your tuition myself." Fortunately, I did get a scholarship that semester, so I didn't have to accept Johansen's incredibly kind offer. I understand that Johansen had helped several students who had financial difficulties.

As a freshman, I took two graduate courses in mathematics, and interesting courses in psychology and philosophy. While on the subject of philosophy, here is another quiz:

Know Your Philosophers

  (1) What ancient Greek philosopher moved very slowly?
  (2) What ancient Greek philosopher had a powerful punch?
  (3) What ancient Greek philosopher exhibited much curiosity?
  (4) What ancient Greek philosopher had fun with part of his foot?
  (5) What ancient Greek philosopher, mathematician and musician was snake-like?
  (6) What 17th century philosopher resembled a vehicle moving in sunlight?
  (7) What philosopher had canine characteristics?
  (8) What philosopher was high up in the scale of evolution?
  (9) What philosopher was very well read?
  (10) What philosopher couldn't do things?
  (11) What philosopher exhibited kitchenware every hour?
  (12) Why can't the philosophers Kant and Reid ever understand each other?
  (13) What philosopher resembled a certain species of bird covered with dry grass?
  (14) What philosopher once physically assaulted a prison guard? [This is one of my favorites!]
  (15) What philosopher, mathematician and theologian resembled a certain brand of cookies?

(16) What philosopher and logician had a good place to swim?

(17) What philosopher and social evolutionist, had he been a woman, would never have gotten married?

(18) What philosopher and biologist won a lot of black substance?

(19) What philosopher would dig on top of a high mountain?

(20) What philosopher resembled a funny stone?

(21) What philosopher who taught at Columbia University resembled a certain kind of food particularly liked by the Jews?

(22) What Jewish philosopher who taught at CCNY was Zen-like?

(23) What Harvard philosopher and psychologist would pacify wild beasts?

(24) What Harvard philosopher and psychologist resembled a certain river in England?

(25) What 19th and 20th century American objective idealist Harvard philosopher resembles a very expensive automobile?

(26) What Harvard philosopher was always pawning things?

(27) What Harvard professor would shock people?

(28) What Harvard philosopher would happily make love?

(29) What philosopher had very strong biceps?

(30) What philosopher swore a lot?

(31) What philosopher was always fussing around?

(32) What philosopher always hurried?

(33) What philosopher was quite mysterious?

(34) What philosopher liked to fight?

(35) What philosopher resembled the sound of leaves in the wind?

HINT: The philosophers 29–35 are all the same. #35 should really give it away!

BONUS PROBLEM: What does Raymond Smullyan become when he is in a logical mood?

Answers

(1) Aris-Turtle

(2) Sock-raties

(3) Epi-curious

(4) Play-toe

(5) Py(thon)agorous

(6) Day-Cart

(7) George Bark-ley

(8) David Human

(9) Thomas Reid

(10) Immanuel Can't

(11) Show-Pan-Hour

(12) Because Kant can't read Reid and Reid can't read Kant.

(13) W.F. Hay-Gull

(14) Soren Kick-a-Guard

(15) Sir Isaac Fig Newton

(16) George Pool

(17) Herbert Spinster

(18) Charles Tar-Win

(19) Martin High-Digger

(20) Ludwig Wit-gen-Stein

(21) Ernest Bagel

(22) Morris Koan

(23) William Tames

(24) William Thames

(25) Josiah Rolls Royce

(26) William Hocking

(27) William Shocking

(28) Arthur Lovejoy

(29) Bertrand Muscle

(30) Bertrand Cussel

(31) Bertrand Fussel

(32) Bertrand Hustle

(33) Bertrand Puzzle

(34) Bertrand Tussle

(35) Bertrand Rustle

ANSWER TO BONUS PROBLEM: When Raymond Smullyan is in a logical mood, he becomes a Boolean Smoolean.

More Philosophers

(1) What philosopher and theologian bathed frequently?

(2) What famous logician never needed to bathe?

(3) What philosopher liked to fall asleep in an automobile?

(4) What ancient Greek philosopher was able to philosophize even after he was dead?

(5) What philosopher was never truthful?

(6) What philosopher was lost without a key?

(7) What utilitarian philosopher had difficulty walking?

(8) What philosopher had part of his anatomy rotating?

(9) What famous Jewish philosopher brings me much money and good contracts?

(10) What French philosopher used a needle?

(11) What Dutch philosopher uses a pin?

(12) What French philosopher resembled a bad part of a tree?

(13) What German philosopher is related to part of one's leg?

(14) What philosopher is related to a German city?

(15) What German philosopher once pulled a lady out of the water?

(16) What pragmatic philosopher got things done?

(17) What pragmatic philosopher was sick and teary-eyed?

(18) What Ancient Greek philosopher healed after injuring his leg?

(19) What famous philosopher and logician resembles a certain piece of female apparel?

(20) What two philosophers were very helpful to those suffering from malaria?

(21) What contemporary logician's name is the modern version of the name of a medieval philosopher?

(22) What philosopher suffered a lot?

(23) What British philosopher was always assenting?

(24) What American mathematician, logician and philosopher would have made the best handbag?

(25) What British philosopher always got people worked up?

(26) What British lady philosopher always had her hair very neat?

(27) What lady philosopher was always ranting and raving?

(28) What lady philosopher was at her best in her private dressing room?

(29) What ancient Greek lady philosopher and mathematician was well paid?

Answers

(1) Karl Bath

(2) Steven Cole Cleaney

(3) Rudolph Car-Nap

(4) Die-Ogenies

(5) Gottfried Lie-bnetz

(6) John Lock

(7) Thomas Hobbles

(8) Spin-Nose-A

(9) My-Money-Deeds

(10) Rou-Sew

(11) Pin-oza

(12) Mal-Branche

(13) Friedrich Knee-chee

(14) Isaiah Berlin

(15) Johann Gottlieb Fished-Her

(16) John Dewey

(17) Jaundice Do-ey

(18) Epi-Mend-Knees

(19) Kurt Girdle

(20) Thomas Aquine-nine and Willard Orman van Quine-nine

(21) Dana Scott is the modern version of Duns Scotus

(22) Thomas Payne

(23) A. J. Ayer

(24) Charles Sanders Purse

(25) Gilbert Rile

(26) Elizabeth Anse-Combe

(27) Ayn Rant

(28) Simone de Boudoir

(29) High-Pay-tia

Still More Philosophers

(1) What 18th century philosopher and writer was not the greatest?

(2) What 18th century German philosopher and writer rented things out?

40

(3) What early 20th century French philosopher was the offspring of a city?

(4) What philosopher provides anesthesia?

(5) What philosopher has failed?

(6) What philosopher and logician resembles a beautiful bird?

(7) What logician is like a blackened article worn on one's foot in the snow?

(8) What logician goes well with rice?

(9) What writer with philosophical tendencies was untamable?

(10) What American author who sometimes waxed philosophical was related to part of one's foot?

(11) What philosopher of Columbia University is related to a Divine Comedy?

(12) What Catholic philosopher took care of ships?

(13) What Friedrich, a 19th century German philosopher, was worth about 25 cents?

(14) What 20th century American moral and political Harvard philosopher resembles a very expensive automobile?

(15) What philosopher who was at the University of Chicago was very good at arithmetic?

(16) What 18th century philosopher and theologian was lacking in color?

(17) What philosopher is ape-like?

(18) What medieval philosopher struck hard?

(19) What ancient Greek philosopher carries us through the air?

(20) What Italian philosopher was not very subtle?

(21) What philosopher loves nothing?

(22) What philosopher resounds?

(23) What philosopher was very good?

(24) What philosopher and mathematician was very bad?

(25) What philosopher was very clever?

(26) What philosopher was very stupid?

(27) What philosopher and logician is good at golf?

(28) What philosopher and logician is very church-like?

(29) What Russian author and philosopher spun false yarns?

(30) What 20th century philosopher always had a cold?

(31) What two philosophers provide good seasoning?

(32) What lady philosopher had canine tendencies?

(33) What philosopher resembles a lion?

(34) What 18th century French satirical writer and philosopher was like electrically-charged atmosphere?

(35) What philosopher truly exists?

(36) What philosopher had keen ears?

(37) What 18th century German philosopher took good care of his sheep?

(38) What 20th century logician takes good care of his sheep?

(39) What philosopher is related to one?

(40) What 12th century Spanish philosopher was always moving a boat?

(41) What ancient Greek philosopher would offer things then not give them?

(42) What ancient Greek philosopher would tell tales about other philosophers?

(43) What ancient Greek philosopher choked others?

(44) What ancient Greek philosopher philosophized even as a baby?

Answers

(1) Gotthold Ephraim Lessing

(2) Gotthold Ephraim Leasing

(3) Henri Berg-son

(4) Michael Novac-aine

(5) Ludwig von Miss

(6) Moses Schoenfinkel

(7) Alfred Tar-Ski

(8) Haskell Curry

(9) Oscar Wilde

(10) Edgar Allan Toe

(11) Arthur Dante (Danto)

(12) Jacques Maritime

(13) Friedrich Shilling

(14) John Rowles Royce

(15) Mortimer Adder

(16) William Paley

(17) Noam Chimsky

(18) William of Sockum

(19) Hera-Glide-Us

(20) Benedetto Gross (Croce)

(21) Theodore Adore-no

(22) Umberto Echo

(23) Nelson Good-man

(24) Herman Vile

(25) J. J. Smart

(26)   Michael Dummet

(27) Hilary Put-nam

(28) Graham Priest

(29) Leo Tall-Story

(30) Walter Cough-Man (Kauffman)

(31) Karl Pepper and Solomon Pepperman (Feferman)

(32) Susan Wolf

(33) Richard Roar-ti

(34) Volt-Air

(35) Jonathan-Is-Real (Israel)

(36) Ernest Cass-Ear-ra

(37) Johann Gottfried Herder

(38) John Sheperdson

(39) Victor Cousin

(40) Ever-Rows

(41) Socra-Tease

(42) Aris-tattle

(43) Aris-throttle

(44) Aris-TOT-le

5 | University of Chicago, Music, Magic

After three semesters at the University of Wisconsin, I transferred to the University of Chicago, and a year or two later I went back to New York. There, for the first time, I did magic professionally. My first job was at a charming night club in Greenwich Village known as the "Salle de Champagne". I did only close-up magic at tables and earned my money exclusively from tips. It was there that I met my first wife. After some time, she persuaded me to go back to the University of Chicago, which I did.

An amusing incident at Chicago: I was required to take an English examination in lieu of taking an English course. One of the main questions on the exam was to describe some novel I had read. I was in a mischievous mood, and I simply came up with a title for a novel and then made up a story to go with it. The upshot is that I passed the exam.

I took courses in both the mathematics department and the philosophy department. The courses that benefited me the most, by far, were in the philosophy department — particularly the three courses I took from the philosopher Rudolf Carnap, who later was most helpful to me in my career.

I very recently received an invitation from David Edmonds (the creator, along with fellow philosopher Nigel Warburton, of the Internet podcast series found at philosophybites.com), to recount my memories of Carnap, and here is the letter I wrote in reply:

Dear David,

Here are some of my memories of Carnap:

... I first knew Professor Carnap when I was an undergraduate at the University of Chicago, majoring in mathematics. Virtually none of the mathematics courses were really of any help to me, since my interests were only in mathematical logic. But the three courses I took with Carnap were enormously beneficial! The first course was in mathematical logic, the second was in syntax, and the third was a seminar in syntax and semantics. The students were graded just on the basis of term papers. For the first course I wrote a paper on a new decision procedure for monadic first-order logic, which yielded, as corollaries, two decision procedures of Quine. Professor Carnap returned the paper with a grade of A and wrote that I should send a copy of my paper to Quine, which I did.

To digress for a moment, after a while, Quine returned my paper, saying, "I appreciated your matrix approach to the problem, but I don't think your paper has the makings of a publishable paper. Considering your unusual insight as to what makes quantification theory tick, I think you should just tinker for a while. Something more strikingly original might well come out!" Throughout my paper, I kept misspelling the word "procedures", and Quine, with his delightful sense of humor, wrote, "You have misspelled *procedures* with admirable consistency!"

Coming back to Carnap, for the next two courses, I failed to hand in papers at the end of the course, and he rightly gave me *incompletes* in both courses. Several months after the end of the last course, I visited Carnap in Princeton and told him, "I have now completed two excellent papers, which I will soon send you." [Yes, I was sufficiently immodest to classify my papers as "excellent"!] After returning home and sending him the two papers, he returned them with A's on both, and evidently remembered my calling my papers "excellent", because he wrote me, "Your self-evaluation was correct." He added, "I wrote to the registrar to change your incompletes to A's. For me, of course, you should get an A⁺ in both courses, but to the registrar, no such subtle distinction is possible." On one of the two papers he wrote, "I think you should consider publishing this in the *Journal of Symbolic Logic*." On the other he wrote, "Again, I strongly urge you to publish this!"

Well, I did publish the first one in the *Journal of Symbolic Logic* under the title "Languages in which Self-Reference is Possible," and it was very well received. As for the second, I published parts of it in papers and parts in my book *Theory of Formal Systems*.

Professor Carnap has been helpful to me in so many ways! When he was in Princeton, he showed my term paper "Languages in which Self-Reference is Possible" to Gödel. Some years later, when I was a graduate student at Princeton, I met Gödel, who congratulated me on my paper, which he thought was my Ph.D. thesis! This is so strange. Although I knew my paper was good, it was hardly enough for a Ph.D. thesis!

Coming back to my student days at Chicago, a fellow student in Carnap's class was Stanley Tennenbaum, who subsequently published two very important results in mathematical logic. The three of us formed a quite friendly circle. Stanley and I frequently walked Carnap home after class. For Carnap's 60th birthday, Stanley arranged a birthday party for him at Stanley's house. Carnap was fond of classical music, and at the party I played the piano for him. When I finished, he said, "If I could play like that, I would be at the piano for twelve hours a day!"

After the party at Stanley's house, Stanley and I and Carnap went to Carnap's place, and there I gave him my proof of the existence of God. Now, I was a magician at the time, and my proof went like this: I started with a deck of cards, and took out a black one — the Queen of Spades, which I *apparently* placed face down on the table, and said, "You know that of two propositions p and q, if p is true, then either p or q is true." He assented. I then said, "Let p be the proposition that the card is black and let q be the proposition that God exists. Since the card is in fact black, then *either* the card is black *or* God exists." He assented. I then turned over the card, which to everyone's surprise was red! I then said, "Since the card is not black, then God exists!" His response was delightful. He said, "Ah, yes! Proof by legerdemain. Same as the theologians use!"

On another occasion, when someone else showed him a trick, which consisted of an impossible-looking physical phenomenon, Carnap said, "Oh, Heavens! I did not think this could happen in *any* possible world, let alone this one!"

Carnap suffered greatly from a bad back (as do I). When I visited him in Princeton, he was lying in bed the whole time with a bad back. At one point, I suggested to him the possibility that the problem might be psychogenic, and suggested that he might consult a psychotherapist. He followed my suggestion, and did so. Some years later, I met the psychotherapist, who spoke to me very highly of Carnap.

I already told you that Carnap was helpful to me in many ways. Indeed, he may well have changed the course of my life! It happened as follows.

Some months after visiting Carnap in Princeton, when I was back in Chicago working as a magician, I one day received a phone call from Professor John Kemeny, who was the chairman of the mathematics department at Dartmouth College in Hanover, New Hampshire. He told me that the department needed a replacement for a math instructor, and that when he was in Princeton, Carnap had recommended me as a "brilliant mathematician" who would be well-suited for the job. Kemeny asked me if I would come to Dartmouth at their expense for an interview, which I did. I went to Dartmouth, got the job, and taught at Dartmouth for two years. So there I was, an instructor at Dartmouth, without even a Bachelor's degree or even a high school diploma! Later the University of Chicago gave me a Bachelor's degree based partly on courses I had never taken, but had taught at Dartmouth.

At the end of the two years, it was John Kemeny who was very helpful in getting me accepted as a graduate student at Princeton. Here Carnap was again helpful, since, at my request, he wrote a letter of recommendation for me to Princeton. In answer to my request for the letter, he wrote me, "Of course I will write a letter for you. You know I think highly of your abilities and wish to be helpful where I can."

If I had not known Carnap, I surely would never have gotten a job at Dartmouth, and probably would never have gone to Princeton, and it is doubtful that I ever would have gotten a Ph.D. in mathematics anywhere else. How would I have ended up? Most likely as a concert pianist or as a magician!

The last time I saw Carnap was at Princeton, shortly after I had completed my Ph.D. thesis. His last words to me were, "I understand you wrote a brilliant thesis."

...

David, I realize that I was extremely immodest in telling you all this, since it puts me in such a favorable light! Well, the sad fact is that I am unfortunately incurably immodest. I really can't help it; it's genetic. Mark Twain once said, "I was born modest, but it didn't last long." Well, I can accordingly say, "I was born immodest, and it has lasted till this day!"

Best regards,

Raymond

P.S. I suddenly remember two other incidents about Carnap. Once in class, instead of saying, "When I wrote *Logical Syntax of Language*," he mistakenly said, "When I wrote *Principia Mathematica*." When I later pointed this out to him, he was amused and referred to it as a Freudian slip. He said, "I guess in my subconscious, I wish I had written *Principia Mathematica*."

On another occasion, he was talking in class about Gödel. He was one of the readers of Gödel's Ph.D. thesis on the completeness of first-order logic. Carnap said, "That was the shortest thesis I ever read in my life!"

Carnap also told us that he had once suggested to Gödel that they write a joint paper. "This might help you make a name for yourself," Carnap had suggested. He said Gödel drew himself up stiffly and said, "*I* will make a name for myself."

* * *

Incidentally, I am not only immodest, but generally incorrigible! Indeed my epitaph should be

IN LIFE HE WAS INCORRIGIBLE.
IN DEATH HE'S EVEN WORSE!

Coming back to my Chicago days, it was in Carnap's classes that I first heard of Gödel's theorem, which we recall is that for any mathematical system of sufficient strength in which only true sentences are proved, there must be sentences, which, though true, are not provable in the system. I have thought of many way of explaining the essential idea behind Gödel's proof to the general public. Here are some of them.

Consider the following sentence:

(A) THIS SENTENCE CAN NEVER BE PROVED.

I am assuming that only true sentences can be proved. Now comes a little paradox: Could sentence (A) be false? No, because if it were, then contrary to what it says, it *could* be proved. But this contradicts the given condition that only true sentences can be proved. Thus, the sentence cannot be false; it must be true.

Now I have just proved that the sentence is true. Since it is true, then as it correctly says, it can never be proved. So how come I just proved it? What is wrong with the above reasoning? The answer is that the general notion of *proof* is not well defined. Yes, for particular mathematical systems, the notion of *proof within the system* is well defined, and one important purpose of the field known as mathematical logic is to make precise the notions of proof within mathematical systems. Now, let us consider a system — call it "System S" — consisting of English sentences in which the notion of proof *within System S* is well defined and in which all provable sentences are true. Now consider the following sentence:

(B) THIS SENTENCE IS NOT PROVABLE IN SYSTEM S.

The paradox disappears. Instead, we have an interesting truth: If the sentence were false, then, contrary to what it says, it would be provable in System S, but only true sentences are provable in System S. Hence, the sentence can't be false and must be true. Since it is true, then like it says, it is not provable in System S. It corresponds to Gödel's famous sentence that asserts its own non-provability in the system in question.

Here is another illustration in the form of a problem: A logician named Hal visits an island of knights and knaves. Hal knows the following truths about the island: that every inhabitant of the island is either a knight or a knave, and that knights make only true statements, and knaves make only false ones. As a logician Hal is always accurate in his reasoning, so that, given true premises, everything he can prove from them is true.

Hal meets a native A who says, "You cannot prove that I am a knight." The problem is to show that A is a knight, but that Hal cannot ever prove that he is.

Here is our solution: If A is a knave, his statement is false, which means that Hal *can* prove that A is a knight, hence prove something that is false (under the assumption that A is a knave), which would be contrary to the given condition that Hal is completely accurate (i.e. that everything Hal can prove is true)! Thus A can't be a knave; he must be a knight. It further follows that since A is a knight, his statement is true, which means that Hal cannot prove that A is a knight. And so the upshot is that A really is a knight, but the logician cannot prove that he is!

However, doesn't this raise a paradox? What is there to prevent the logician from going through the same reasoning we went through to prove that A is a knight? If he could do that, Hal would have *proved* that A is a knight, which would then make A's statement false, proving to Hal that A is really a knave, contrary to Hal's proof that A is a knight! Hal would have proved both that A is a knight and not a knight, making him inconsistent, and since one of the two statements must be false, also inaccurate.

Well, what is there to prevent this from happening? Is there something you and I know which the logician Hal doesn't know and cannot prove? Yes, there is! I told you that Hal is completely accurate in his proofs, but I never told you that Hal knew this, or could prove this! In fact, if he doesn't know that he is always accurate, he can't take the very first step in the argument in our solution above that A is a knight (the part of the proof involving proof by contradiction that A can't be a knave), since our solution relies on that assumption. And we have seen that if Hal *could* prove that everything he could prove is true, then he could prove inconsistent and inaccurate statements. [This is somewhat related to Gödel's second incompleteness theorem, which is that for systems of sufficient strength, if they are consistent, that consistency cannot be proved in the system.]

Returning to Gödel's first incompleteness theorem, here is another illustration: Consider a machine that prints out expressions built from the following symbols:

F N R *

Call an expression *printable* if the machine can print it. We assume that the machine is so programmed that any expression it can print will be printed sooner or later. By a *sentence* is meant any expression of one of the following four forms, where "X" stands for any expression whatsoever:

(1) P*X
(2) NP*X
(3) PR*X
(4) NPR*X

The sentences are to be interpreted as follows: P*X is to be interpreted as "X is printable", and is accordingly called *true* if and only if X is printable. The symbol "P" stands for "printable". The symbol "N" stands for "not" and thus NP*X is called true if and only if X is not printable [NP*X can be read as "not printable X", or, in better English, "X is not printable"]. The symbol "R*X" stands for "repeat X", where by the repeat of an expression X is meant XX (X followed by itself), and so PR*X is called *true* if and only if XX is printable. Finally, NPR*X is called *true* if and only if XX is not printable.

The symbol "*" is for punctuation. If it wasn't in the language at all, ambiguities could arise — for example, one could not tell whether PRX meant that RX is printable, or that XX (the repeat of X) is printable. If the former is meant, using the punctuation symbol * it should be written P*RX; if the latter, it should be written PR*X.

We are given that the machine is wholly accurate, in that every sentence it prints out is true. Thus, for example, if PR*X is printable, it is true, which means that XX will be printed out sooner or later. Now suppose X is printable. Does it follow that P*X is printable? Not necessarily. If X is printable, then P*X must be true, but does not mean that P*X must be printable. I never told you that all true sentences are printable. I told you only that all printable sentences are true — in fact there is a true sentence that the machine cannot print! The problem is to find one. Can the reader do so without reading further?

Well, for any expression "X", the sentence NPR*X is true if and only if the repeat of X is not printable. Since this is true for any X, it is true for the special case that X is the expression NPR*. Thus NPR*NPR* is true if and only if the repeat of NPR* is not printable. But the repeat of NPR* is the very expression NPR*NPR*!

Thus NPR*NPR* is true if and only if NPR*NPR* is not printable. This means that NPR*NPR* is either true and not printable, or printable but not true. The latter alternative is ruled out by the given condition that only true sentences are printable. Hence, the sentence NPR*NPR* is true but, the machine cannot print it. [This corresponds to Gödel's famous sentence that asserts its own non-provability in the system under consideration.]

Obviously, no machine could be accurate if it prints out a sentence that asserts that it is not printable! This is reminiscent of the scene in Romeo and Juliet in which the nurse runs to Juliet and says, "I have no breath!" at which Juliet says, "How can you have no breath when you have breath to say 'I have no breath'?"

There is something else that is quite interesting about this machine: It is possible to construct two sentences X and Y such that one of the two is true but not printable, but there is no way of knowing which it is! The idea is to construct them such that X asserts that Y is printable, and Y asserts that X is not printable — that is, we want X to be true if and only if Y is printable, and want Y to be true if and only if X is not printable. It then follows that one of those two sentences must be true but not printable (assuming, of course, that all printable sentences are true), but there is no way of telling which one it is. Why is that so? Well, here is the argument:

Either X is true or it isn't. Let us first consider the case that it is. Then Y is printable, as X correctly says. Since Y is printable, it must be true. Hence, as it correctly says, X is not printable. Thus, if X is true, then X is not printable.

Now let us consider the case that X is not true. Then X is not printable (since only true sentences are printable). Therefore Y, which asserts that X is not printable, is true. But also, since X asserts that Y is printable, and X is false, then Y is really not printable. And so in this case, Y is true but not printable.

In summary, if X is true, then X is true but not printable, whereas if X is false, then it is Y that is true but not printable. There is no way of knowing whether X is true or not, hence there is no way of knowing whether it is X or Y which is the one that is true but not printable.

Now, how do we construct such an X and Y from the given symbols? Well, take X to be P*NPR*P*NPR*, and Y to be NPR*P*NPR*. Now X is the expression P*Y, and thus asserts that Y is printable. Y asserts that the repeat of P*NPR* is not printable, but the repeat is X.

There is another solution: Take X to be PR*NP*PR* and Y to be NP*X, which is NP*PR*NP*PR*. X asserts that the repeat of NP*PR* is printable, but this repeat is NP*PR*NP*PR*, which is Y. Thus X asserts that Y is printable. Y, which is NP*X, obviously asserts that X is not printable.

Gödel's construction made me think of the following logic puzzle: I put a penny and a quarter on the table and ask you to make a statement. If the statement is true, then I will give you one of the two coins, not saying which one. If the statement is false, then I won't give you either coin. The problem is this: Which statement could you make such that my only option is to give you the quarter (assuming that I keep my word)? I will give you the answer to this one and to the next three problems after the conclusion of their statements.

There is another statement you could make such that I would have to give you both coins. What statement would work?

There is yet another statement you could make which would make it impossible for me to keep my word. What statement would that be? [This should be pretty obvious!]

Consider the original form of the puzzle, which I have been using for years in teaching my logic classes: I would ask a student what statement would win the quarter. Then one day, I suddenly realized to my horror that I left myself wide open for the student to win any amount — say a million dollars — from me! What statement could the student make such that in order to keep my word, I would have to pay the student a million dollars?

Now for the answers to all four puzzles.

What statement could you make which would force me to give you a quarter? Well, one such statement is: "You will not give me the penny." If the statement is false, then contrary to what it says, I *would* give you the penny, which violates the given condition that I don't give you either coin for a false statement. Therefore, the statement can't be false; it must be true. Since it is true, then like it says, I don't give you the penny. But I must give you one of the two coins for a true statement. Hence, I must give you the quarter.

I said that this puzzle was suggested to me by Gödel's theorem. How? Well, I thought of the penny as corresponding to provability and the quarter as corresponding to truth, and so the statement "You will not give me the penny" corresponds to Gödel's sentence which says "I am not provable".

Next, what statement could you make that would force you to give both coins? One such statement is, "You will either give me both coins, or neither coin." The only way the statement could be false is that I give you one of the two coins but not the other, but I can't give you a coin for a false statement. Therefore, the statement can't be false; it must be true, which means that I really will either give you both coins or neither one, but I can't give you neither one for a true statement. Hence I must give you both.

Next, what statement would make it impossible for me to keep my word? Obviously such a statement is "You will give me neither coin." If I gave you at least one of the coins, that would falsify your statement, so I would be giving you a coin for a false statement. But I couldn't do that because it would be violating what I said I would do. On the other hand, if I give you neither coin, that would make your statement true, however I would have failed to give you a coin for a true statement, so again I would have broken my word.

This puzzle is like that of the classical puzzle of the town which decrees that any stranger entering the town must make a statement. If the statement is true, he is to be sacrificed on the altar of truth, and if his statement is false, he gets sacrificed on the altar of falsehood. What statement could a stranger who enters make which would make it impossible for the inhabitants of the town to carry out this decree? Well, such a statement is "I will be sacrificed on the altar of falsehood."

Finally, what statement could you make that would force me to give you a million dollars? Well, all you need to say is "You won't give me either the penny, or the quarter, or a million dollars." If the statement were true, I would have to give you either the penny or the quarter by the initial rules of the game, but doing so would make it false that I give you neither the penny, nor the quarter, nor a million dollars, and we would have a contradiction. Hence, the statement can't be true and must be false. Since it is false, then the opposite must be the case, which means I give you either the penny or the quarter or a million dollars, but I can't give you either the penny or the quarter for a false statement, hence I must give you a million dollars.

* * *

In the days when I was attending the University of Chicago, I also worked at night as a magician to support myself and my wife. I worked at night clubs as a close-up magician, going from table to table. At one table I came across, sat just one man — the most blasé character I ever met! He just sat there smoking his pipe and showing not the slightest reaction to any of my tricks. I made my tricks better and better, but still not the slightest response. Finally, I did my most spectacular trick, upon which the man took the pipe out of his mouth, angrily slammed the table with his fist and yelled, "It's a trick!"

I am reminded of an incident involving Mark Twain. He was lecturing at some provincial place in Maine or Vermont and got no reaction from his audience. He made his jokes funnier and funnier, but all faces remained completely deadpan … not a single smile. Mark Twain wondered, "Am I losing my touch?" Well, during intermission, he heard an elderly couple discussing his act. The man said to his wife, "Weren't he funny? Weren't he funny? You know, at times I could hardly keep from laughing!"

Speaking of Mark Twain, once he had to give a talk at a banquet. He was very tired at the time, and so, when the time came for his speech, he wearily rose up and said, "Homer is dead. Shakespeare is dead. And I am none too well!"

Speaking of Vermont, many jokes are told. It seems to me that the essence of Vermont humor is that when you ask a Vermonter a question, the answer is correct but highly insufficient! As an example, a Vermont farmer once went to a neighboring farmer and asked, "Lem, what did you give your horse the time it had the Colic?" Lem replied, "Bran and molasses." The farmer went away and returned a week later and said, "Lem, I gave my horse bran and molasses and it died!" Lem replied, "So did mine!"

Another Vermont story is about a man who was driving through Vermont and came to a fork in the road. On one road of the fork was a sign which said "To White River Junction". On the other road of the fork was the very same sign. The driver got out of the car and asked a man standing close to the fork, "Does it make any difference which of the two roads I take?" The man replied, "Not to me, it doesn't."

Actually, many Vermonters have a keen, though wry, sense of humor. A student once told me that he drove past a farm house in Vermont and saw a farmer rocking on a rocking chair on the porch. Being in a mischievous mood, he said to the farmer, "Been rocking that way all your life?" The farmer replied, "Not yet."

President Calvin Coolidge was from Vermont. Many stories were told about him. He spoke very little and was often called "Silent Cal". At a banquet, he was sitting next to a lady, and a half an hour passed without his saying a single word. Finally, the lady said, "Mr. President, I have a bet that I can get more than two words out of you!" Coolidge replied: "You lose!" Now, that surely shows great cleverness, does it not? Another incident is about his coming home from church one Sunday. A neighbor asked him, "What did the minister talk about today?" Coolidge replied, "Sin." The neighbor asked, "What did he have to say about it?" Coolidge replied, "He was agin' it!"

There was a story that Coolidge once visited a farm with some friends. They saw some sheep. One of the friends said, "I see the sheep have been shorn." Coolidge said, "It looks like it from this side!"

There is another incident I heard about Coolidge that increased my respect for him enormously. He once came into his office and found a cat burglar rummaging through his things. The burglar burst into tears and explained that he was desperately looking for money to pay for fare to travel to his mother who was dying. Coolidge reached into his pocket, pulled out some money, and gave it to the burglar, saying, "Please pay me back whenever you can and be careful how you exit here since this place is heavily guarded!"

I am reminded of the following incident. A friend of mine, who is a baker, once went into a famous delicatessen in New York City. She saw an elderly lady slip something into her bag. When she told the owner, he replied, "Oh yes, she is a good customer, but occasionally she steals something." My friend asked him, "Don't you ever speak to her about it?" He replied, "Oh no, that would embarrass her terribly!"

Back to my magical activities in Chicago. People, knowing that I was a magician, often asked me whether I had ever sawn any ladies in half. I would reply, "I have sawn dozens of ladies in half, and I'm learning the second half of the trick now!"

Speaking of magic, let me tell one of the most elegant magic tricks that I know. It was done by Sam Lloyd, one of the greatest puzzle inventors of the world, and also a good magician. It was done aboard ship with his twelve-year-old son. The boy stood blindfolded with his back to the audience. A spectator had his own deck of cards, shuffled them and showed them one by one to Sam Lloyd, and each time the boy correctly identified the card. How was it done? This was at the turn of the century before there were such things as radio signals. Even magicians were baffled. So, how was it done? I will tell you later on, after telling you about four more tricks, whose explanations will also be given then.

Here is a trick done by the legendary Houdini. In full view of the audience, Houdini, who was naked except for a loincloth, stepped into a hot oven with a steak. There was a glass door so that everyone could see what was going on inside. After a while, Houdini came out with a steak that was well cooked, but Houdini was unharmed. The steak that Houdini brought out was really the same one that he took in. How was it done?

Next, a mind-reading trick that was subsequently known as "Jimmy Valentine opens a safe!" In a public square was a safe owned by one of its citizens, who of course knew the combination to open it. When the safe was opened, the lady assistant of the magician entered it, and the door was closed, locking the lady inside. The magician went to the safe and asked the owner to concentrate deeply on the combination. The magician then played with the knob, turning it back and forth, and finally the door opened and the lady made her triumphant exit to the great applause of the audience. How was it done?

It is amazing to what lengths some magicians go to produce certain effects. Here are two examples: Towards the beginning of the 20th century, one afternoon a magician came into a small town. He was allowed to make his preparations at the theatre in which he was to perform in the evening. When the evening came, the audience saw on the stage a table with a closed deck of cards on it, and behind the table was the magician seated on a chair. The magician called for some member of the audience to assist him. A man came up who was not a confederate (the magician knew no one in the audience) and the magician asked him to take the cards out of the case, examine to see that there were 52 cards, all different, take the cards down to the audience, have three people sitting there each select a card and conceal it in his or her pocket or purse, return the other 49 to the case, bring the case up to the stage and put the card case on the table. The magician then named the three missing cards. How it was done, I will soon tell you.

Another trick: At a party, a card was selected, shown around, and returned to the deck, which was then shuffled. A basket of eggs was brought in, an egg was selected, the shell broken, and inside the yolk was a piece of paper on which was written the name of the card. The hardest part of the trick was to get the slip of paper into the yolk of the egg without leaving any telltale marks on the surface of the shell. How was it done?

Now for the explanation of the five tricks: For the Sam Lloyd trick, the son never said a word; Sam Lloyd was a ventriloquist! This was about the most elegant trick that I know. Humorously enough, one elderly gentleman told the father, "You shouldn't strain the boy's mind so much. It's not good for him!"

As for the Houdini in the oven trick, anyone can do it, though I hardly recommend it! The fact is that Houdini, being alive, perspired. Any live person who normally perspires can stand a temperature hot enough to cook a steak. Had Houdini been dead, he also would have been cooked.

Next, the trick of opening the safe. Was it mind reading? Not at all. The lady assistant entered the safe with a concealed screwdriver making it very easy for her to open the safe from the inside. Clever, huh?

Now for the trick in which the magician of the small town named the three missing cards. Well, hidden above the table was an X-ray machine. The deck of cards was a prepared deck. On the back of each card was a spot of metallic salt, impervious to X-rays. Different cards had their spots of metallic salt in different positions. When all 52 cards were in the pack, no X-rays could go through any part of the deck where one of the cards had a metallic salt on it, but if three cards were missing, then there were, so to speak, three holes in the places of spots through which X-rays would now pass, and the dark image of the deck (with white spots on it for 49 cards) was reflected onto a florescent machine hidden at the base of the table, which the magician could see.

Now for the trick of getting a slip of paper into the yolk of an egg. Well, the magician had previously planted the slip of paper into the ovary of the chicken and the egg grew around it!

Coming back to Houdini, I will now tell you some little known facts about him. A particularly delightful incident is about how Houdini was once outwitted by a British police officer. As many of you know, it normally took Houdini about eight minutes to escape from any prison cell. Well, in England, it took him over two hours to escape from a prison cell. Why? Because the police officer didn't lock the door, he only pretended to! Houdini was working on an open door the entire time and none of his tricks worked on an open door! Finally, it dawned on him that the door was open!

In addition to being a lover of magic, I am very fond of clever swindles! Here is one known as the "ten dollar bar swindle". It was done at the turn of the 20th century. This has to be done at a large bar where there are two cash registers that are far apart. The magician goes into the bar, and after entertaining everyone for a while says, "Now I will do the greatest trick of the evening! Can I please borrow a ten dollar bill from the register?" The bartender takes a ten dollar bill from the register and hands it to the magician. The magician then asks the bartender to write his name on the bill, which he does. The magician then says, "I will put the bill in this envelope and burn it." He then puts the bill in an envelope and seals it. There is a slit at the back of the envelope, through which the magician takes out the bill, palms it, and hands it to a confederate. [Nobody knows that he has a confederate.] While the magician makes preparations for burning the envelope, the confederate takes the ten dollar bill way over to the other cash register, orders a drink for a dollar, gives the bartender the ten dollar bill, and gets nine dollars change. Meanwhile, the magician burns the envelope in front of all the spectators and says, "I will now magically restore the bill and make it reappear in that cash register over there!" Someone is sent over to the other register and comes back with the bill, which surely enough has the bartender's signature. Then everyone yells "Bravo!"

I am reminded of a joke which takes place in modern times. A man goes into a bar and orders a beer. After drinking it, he is about to walk out when the bartender says, "Hey, you haven't paid me for the beer!" The man says, "I certainly did." The bartender responds, "I don't recollect you having paid me!" The man answers, "Well, I can't help it if your memory is short lived. And besides, the customer is always right ... isn't that so?" The bartender scratches his head and doesn't know what to do, and the man walks out. He meets a friend and tells him, "You want a free beer?" and explains what he did. The friend goes into the bar and pulls the very same trick. The bartender is furious, but again, doesn't know what to do. The friend walks out and meets another friend, whom we'll call "Mr. Wise Guy", to whom he relates what he has just done. Mr. Wise Guy goes into the bar, orders and drinks a beer. The bartender says, "Do you know what just happened to me? Two guys came into this bar and ordered beers and didn't pay for them. And both claimed that they had paid! Now the next guy who tries a trick like that, I'm just going to twist his head around his neck, see!" Mr. Wise Guy says, "Don't bother me with your troubles. Just give me my change. I want to go." [As I first heard this joke, the first two friends were of the same ethnic group, while the "Wise Guy" was of another.]

Here's another joke of which I am quite fond. A man walked into a bar in Ireland and ordered three beers. This went on for several nights. Finally, the bartender could no longer contain his curiosity and asked, "Why do you always order three?" The man replied, "Well, I have two brothers, one is in America and one is in Australia. We have a pact. When one of us drinks a beer, he also drinks two others in memory of his two brothers." Everyone present thought this very sweet. This went on for several months, but one night, to everyone's dismay, the man came in and ordered only two beers. One customer came over to him to console him for the loss of his brother. The man said, "My two brothers are fine!" When asked why he ordered only two beers, the man said, "I've given up drinking."

Coming back to Houdini, his relation with Conan Doyle was remarkable! This was when Doyle was in his later years and involved with spiritualism and the supernatural in a most crazy way! He kept insisting that the way Houdini escaped from locked trunks was by dematerializing and going out through the key hole! Nothing Houdini could tell him would change his mind. Doyle kept writing him passionate letters begging him not to keep the secret of dematerialization for cheap magic tricks, but to share this important information with the world. Doyle was really a very stubborn man in those days. Once in London he saw a mind-reading act between a husband and wife. After the performance, Doyle went backstage to congratulate the couple on their psychic powers. The husband said, "I'm sorry to disappoint you, Sir Doyle, but we don't have psychic powers. We use signals." Doyle angrily replied, "I'm sure you have psychic powers, whether you realize it or not," and stormed out of the room.

Once Conan Doyle and his wife took Houdini to a medium to contact Houdini's dead mother. The medium went into a trance and out came some words, which impressed the Doyles very much, but Houdini was laughing the whole time. After the séance was over, and the three were out on the street, the Doyles asked Houdini why he had been laughing. Houdini replied, "If that had been my mother, she never knew a word of English. All she knew was Yiddish!"

I am reminded of two of my favorite Jewish jokes. A New Yorker was away from the city for a year, and when he came back, he wanted to see how things were at his favorite Jewish restaurant. To his surprise, a Chinese waiter took his order, and those of others, in perfect Yiddish! Later, the man asked the owner, "How come this Chinese is learning Yiddish?" The owner replied, "Shhhh. He thinks he's learning English."

The other joke is about a rabbi who was being driven to another city, where he had never been before, to give a lecture at a synagogue. The chauffeur was acting very disgruntled. When the rabbi asked him what was bothering him, the chauffeur said, "I work so hard and get so little admiration! You work so little and get so much admiration. It's not fair!" The rabbi said, "If you feel that way, let's change roles!" And so the rabbi sat in the front, and the chauffeur sat in back. When they got to the other city, the people assumed that the one in the back was the rabbi and brought him into the temple. The real rabbi followed, in the role of the chauffeur. In the temple, they asked the chauffeur, whom they thought was the rabbi, a difficult theological question. The chauffeur said, "Oh, that's so obvious, even my chauffeur could answer that!"

Coming back to Houdini, here is an incident that I found very moving! As many of you know, Houdini was constantly exposing fraudulent spiritual mediums. One day Houdini said to his wife Beatrice, "Although I disbelieve spiritualism, I want to be open-minded. And so if I should die before you, if there is any possibility of coming back to you, I promise to try and do so within six months after my death!" Well, he did die before her, and a few months later, on the back cover of the Christmas issue of the British magic magazine called "The Sphinx" was a message from Mrs. Houdini:

> DEAR HARRY,
>
> YOU WERE RIGHT AS USUAL.
>
> YOU DIDN'T COME BACK.

To digress for a moment, I am reminded of an incident about Thomas Paine, who escaped execution while in Paris in a most remarkable manner! He was arrested and sentenced to be executed. He was confined in a cell which was one of a whole line of cells. Not all the prisoners in the cells were to be executed, and a guard came by to put a chalk mark on the outside of the doors of those who were to be executed the next day. The door of the cell of Thomas Paine happened to be open at the time the guard came by, who carelessly put a mark on the *inside* of the door!

6 | Dartmouth, Princeton

Coming back to my own life now: As you have already read in the letter I included about Rudolph Carnap at the beginning of Chapter 5, one day towards the end of that period in Chicago I received an unexpected phone call from John Kemeny, the chairman of the mathematics department of Dartmouth College, which led to my obtaining a job at Dartmouth. The reader might find amusing some further details about my being accepted for that position and the consequences it had in terms of my finally obtaining an undergraduate degree: John Kemeny and one other professor wanted me for the job after the interview I had with them, but first Kemeny had to get permission from the Dean. Later, Kemeny said that he had told the Dean that the eminent philosopher and logician Rudolph Carnap highly recommended a very talented mathematician, Raymond Smullyan, but that Smullyan did not have a degree. The Dean replied, "Well, maybe he can work on his Ph.D. while he is teaching." Kemeny responded, "I don't think you understand. He doesn't even have a Bachelor's Degree." There was a long pause. The Dean finally said, "Well, if you believe he is competent, then hire him." The department was evidently pleased with me, since they hired me for a second year. It was then that the University of Chicago gave me a Bachelor's Degree by accepting the courses I had taught at Dartmouth as if I had taken them at Chicago or another reputable university. I understand there was quite a fight over me between the faculty and administration at Chicago. The faculty thought that I should get the degree, but the administration thought that I should not. Fortunately for me, the faculty won.

The chairman, John Kemeny, was also a logician, and he later became president of Dartmouth College. He had formerly been an assistant to Albert Einstein, and he told me an interesting incident about Einstein and Kurt Gödel. Einstein and Gödel had offices opposite each other in a hall of The Institute for Advanced Study. At the time Kemeny was there, Gödel was working on some strange physical theories about universes which were non-existent, but logically possible. One day Kemeny asked Gödel what Einstein thought of the work. Gödel replied that he didn't know, since he had never met Einstein. Kemeny thought that was remarkable, since they had had opposite offices for over a year, and yet had never spoken to each other! Kemeny immediately took Gödel into Einstein's office and introduced them to each other. The two became close friends, and often walked out of the building together, arm in arm.

Here is another interesting story about Einstein and Gödel. After Gödel was in the U.S. for some time, he had to take the examination for citizenship. Einstein and Von Neumann drove him to the place of the examination. While they were en route, Gödel told them that he would tell the committee that he found an inconsistency in the American Constitution. Einstein said, "For heaven sakes, don't do that! If you show your superiority to the committee members, you will never get your citizenship!" Fortunately, Gödel took Einstein's advice.

I must now tell you something very funny. I came across a grade school textbook about arithmetic in which Gödel was mentioned. To be precise, the book contained a photograph under which appeared the name KURT GÖDEL. But the photograph was actually of Einstein! I can understand that many people wouldn't know what Gödel looked like, but I am amazed that someone wouldn't know what Einstein looked like!

I will tell you more about Einstein later on.

It was during my second year at Dartmouth that my first wife and I separated and later got divorced. At the end of that year, I was accepted as a graduate student at Princeton University, largely on the recommendations of Carnap, Kemeny and some mathematicians at the University of Chicago. At Princeton University I studied with two of the foremost logicians — Alonzo Church and Steve Kleene. Officially, Church was my thesis advisor, but most of my work was done during the year that Church was away. The mathematician who had the most influence on me, though, was the brilliant logician John Myhill, who was then at The Institute for Advanced Study, also in the town of Princeton. During my stay at Princeton, I got particularly interested in Gödel's theorem, and in a related result of Alfred Tarski, which roughly speaking is that in the systems investigated by Gödel, truth within the system is not definable in the system. I will now illustrate these results.

The mathematical systems under consideration have the following features. In describing these features, we will use the word *number* to mean a positive whole number or zero — one of the numbers 0, 1, 2, 3, ..., n, ... Among the sets of expressions of the mathematical systems under consideration is a well-defined set of expressions whose members are called *sentences*, and another well-defined set of expressions called *predicates*. Each predicate is the *name* of a set of numbers. For each predicate H and each number n, there is a sentence denoted $H(n)$, which is interpreted to mean that n belongs to the set named by H, and is accordingly called *true* if and only if n really does belong to the set named by H. Every sentence of the system is of the form $H(n)$, for some predicate H and number n.

Every expression of the system is assigned a number called the *Gödel number* of the expression. If a number n is the Gödel number of a sentence, then we call n a *sentence number*, and we then let S_n be that sentence whose Gödel number is n. If n is the Gödel number of a predicate, then we will call n a *predicate number*, and we then let H_n be the predicate whose Gödel number is n. For any predicate H whose Gödel number is h, by the *diagonalization* of H is meant the sentence $H(h)$. Thus for any predicate number n, the diagonalization of H_n is the sentence $H_n(n)$. To each number n we assign a number denoted n^* such that if n is a predicate number, then n^* is the Gödel number of the sentence $H_n(n)$ (the diagonalization of H_n).

The following notation will prove convenient, since it will enable us to avoid constant reference to the Gödel numbering: For any predicate H and any expression X, by $H[X]$ we shall mean $H(x)$, where x is the Gödel number of X. In particular, for any predicate H, the expression $H[H]$ is the diagonalization of H.

Given a set $\mathfrak{S}$ of *expressions* of the system, we will say that a predicate H *defines* $\mathfrak{S}$ to mean that H *names* the set of Gödel numbers of the elements of $\mathfrak{S}$. Thus H *defines* $\mathfrak{S}$ is to say that for every expression X, the sentence $H[X]$ is true if and only if X is a member of $\mathfrak{S}$.

Two sentences are called *equivalent* if they are either both true or both not true (i.e. both false). Now, the systems under investigation satisfy the following two conditions:

C_1: To each predicate H is assigned a predicate denoted $H^\#$ called the *diagonalizer* of H, such that for every predicate number n, the sentence $H^\#(n)$ is equivalent to $H(n^*)$.

C_2: To each predicate H is assigned a predicate $\overline{H}$, which we will call the *negation* of H, such that for every number n, the sentence $\overline{H}(n)$ is true if and only if $H(n)$ is not true.

Let us note that condition C_1 implies that for any predicates H and K, the sentence $H^\#[K]$ is equivalent to $H[K[K]]$, because K has some Gödel number k, and thus $H^\#(k)$ is equivalent to $H(k^*)$, but $H^\#(k)$ is the sentence $H^\#[K]$, and since k^* is the Gödel number of $K[K]$ (which is $K(k)$), then $H(k^*)$ is the sentence $H[K[K]]$.

From just conditions 1 and 2 comes something that I hope will surprise the reader:

Theorem T (Tarski's Theorem in Miniature). *The set of true sentences of such a system is not definable in the system.*

We recall that this means that the set of Gödel numbers of the true sentences of the system is not nameable by any predicate in the system.

An important principle behind the proof of Theorem T, which is also of considerable interest in its own right, is the following: A sentence S is called a *fixed point* of a predicate H if the sentence S is equivalent to $H[S]$. [Fixed points are clearly related to the subject of self-reference: Suppose S is a fixed point of H. Let n be the Gödel number of S. Then S_n, being equivalent to $H(n)$, can be thought of as asserting that its own Gödel number is in the set named by H.]

From just condition C_1 follows:

Theorem F (Fixed Point Theorem). *Every predicate H has a fixed point.*

Would the reader like to try proving the Fixed Point Theorem before reading further? Hint: Show that the diagonalization of the diagonalizer of H is a fixed point of H.

Well, here is the proof of Theorem F. As we have already noted, for any predicates H and K, the sentence $H^*[K]$ is equivalent to $H[K[K]]$. Since this holds for every K, it holds in the case that K is the predicate H^*. Thus $H^*[H^*]$ is equivalent to $H[H^*[H^*]]$, which means that $H^*[H^*]$ is a fixed point of H.

Now for the proof of Theorem T. A predicate H is called a *truth predicate* if it defines the set of true sentences (i.e. if it names the set of Gödel numbers of the true sentences). To prove Theorem T, we must show that no truth predicate can exist in any mathematical system satisfying our assumptions. Well, if H is a truth predicate, then for every expression X, the sentence $H[X]$ is true if and only if X is a true sentence, and so for every *sentence S*, the sentence $H[S]$ is true if and only if S is true, and thus if and only if S is a fixed point of H. Thus, if H is a truth predicate, then *every* sentence is a fixed point of H! Therefore, to show that H is not a truth predicate, it suffices to show that there is at least one sentence which is *not* a fixed point of H. Such a sentence might aptly be called a *witness* that H is not a truth predicate. Well, by the Fixed Point Theorem, there is a fixed point S for $\overline{H}$ (the negation of H). Thus S is equivalent to $\overline{H}[S]$, so that S cannot be equivalent to $H[S]$ (for if it were, then $\overline{H}[S]$ would be equivalent to $H[S]$, which is not possible). Thus any fixed point of $\overline{H}$ is a witness that H is not a truth predicate.

We have seen in the proof of Theorem F that a specific fixed point of $\overline{H}$ is $\overline{H}^*[\overline{H}^*]$ [Incidentally, another fixed point of $\overline{H}$ is $H^*[H^*]$, as the reader can verify.] This proves Theorem T.

Next, the systems under consideration also have the following feature. There is a set of sentences called *provable sentences*, such that the following condition holds:

C_3: The set of provable sentence *is* definable.

The system is called *correct* if only true sentences are provable. We will assume that the systems under consideration are correct, i.e. that every provable sentence is true.

Since the set of provable sentences is definable, and the set of true sentences is not (by Theorem T), then the two sets don't coincide, which means that either some true sentence is not provable, or some provable sentence is not true. The latter alternative is ruled out by our assumption that our system is correct and only true sentences are provable. Therefore, we get Gödel's result that some true sentence is not provable. Such a sentence we will call a *Gödel sentence*. Such a sentence can actually be displayed: The set of provable sentences is defined by some predicate P. Well, any fixed point S of $\overline{P}$ is a Gödel sentence (because $P[S]$ is true if and only if S is provable, hence $\overline{P}[S]$ is true if and only if S is not provable, and so if S is equivalent to $\overline{P}[S]$, then S is true if and only if S is not provable).

We have seen from the proof of Theorem F that $\overline{P}*[\overline{P}*]$ is a fixed point of $\overline{P}$, and so $\overline{P}*[\overline{P}*]$ is a Gödel sentence: It is true but not provable in the system (under the assumption that the system is correct).

I must now tell you an amusing incident. I have a friend Goodwin Sammel, a musician who is also interested in mathematics. When he heard about Gödel's theorem, he came up with the following exchange:

A: It's true!
B: It's not!
A: Yes, it is!
B: It couldn't be!
A: It *is* true!
B: Prove it!
A: Oh, it can't be proved, but nevertheless it's true.
B: Now, just a minute. How can you say it's true if it can't be proved?
A: Oh, there are certain things that are true even though they can't be proved.
B: That's not true!
A: Yes, it is. Gödel *proved* that there are certain things that are true, but cannot be proved.

B: That's not true!

A: It certainly is!

B: It couldn't be true, and even if it were true, it could never be proved.

* * *

Now let me tell you how I got into publishing books (of which twenty-six are now on Amazon, and three, possibly four, are on their way).

I had already published some papers before I entered graduate school at Princeton. My doctoral thesis was entitled "Theory of Formal Systems", and shortly after receiving my Ph.D., the thesis was published as a book by Princeton University Press. This was my first published book.

I already told you that in my high school days, I started composing a series of chess problems of an unusual sort. I later found out that the method that had to be used to solve them was known as "retrograde analysis". Unlike the usual type of chess problems which is of the form, "White to play and mate in so many moves", in retrograde problems a position is shown and the problem is to delve into the past of the game and determine how the position ever arose. The usual type of retrograde question is "What was the last move?" or "Can White castle?" In the retrograde problems I made, the questions were more colorful. For example, in one of my best problems called "The mystery of the missing piece" I tell a story that one of the chess pieces got lost, and a coin is used as a substitute for it. A position is shown in which a coin sits on one of the squares of the board, and the problem is to determine what piece the coin represents. In another of my problems, I show a position in which there are three white knights on the board. Obviously, one of them must be promoted, and the problem is to determine which one.

At the time I started composing retrograde problems, I did not know that such a field already existed. Retrograde analysis was then hardly known in America, although it was somewhat known in Europe. Since there were no American periodicals I knew that published retrograde problems, my problems remained unpublished at the time. Then it occurred to me that this kind of problem would be ideally suited to incorporate into stories. Inspired by Lewis Carroll, I made stories corresponding to my chess problems such that the chess pieces themselves would play the part of the main characters. For some reason or other, the Arabian Nights leaped to mind as the setting: Haroun El Rashid as the White King, his grand vizier as the White King's bishop, and so forth. At that time I wrote an Arabian Nights chess puzzle manuscript and planned someday to make a whole book of them. Well, years later, when I was in Princeton, I showed many of these problems to several of my fellow graduate students, as well as to a mathematical logician who was at the nearby Institute for Advanced Study at the time. One of these problems (one of my best) was entitled "Where is the White King?" It consisted of a position in which the White King could not be seen on the board, because he had made himself invisible. The problem was to determine on which square he stood. I showed the problem to my office mate, and he asked me why I didn't publish it before someone else did. I laughingly replied, "Why would anybody do that?"

Well, a few weeks later, a logician from the Institute came into my office and said, "Hey, Smullyan, how come someone published your problem in the Manchester Guardian without crediting you as the author?" [The Manchester Guardian is a British newspaper.] Well, he showed me the paper, and I saw to my amazement that the problem was submitted by the father of my office mate! When I later confronted my office mate and asked him how that happened, he said, "Oh yes, I showed the problem to my father, who has had frequent correspondence with the chess editor of the Manchester Guardian, and he sent him your problem, with the comment that instead of the usual type of problem, he should publish a problem like this!"

The father (who happened to be a prominent political figure at the time) never claimed to be the author of the problem, but merely failed to state who the author was. When I expressed my disappointment that my authorship had not been acknowledged, my office mate told me that he would speak to his father about this. A couple of weeks later, I received a very nice letter from the chess editor of the Manchester Guardian expressing regret that he had not known that I was the author of this "delightful work", and assuring me that my authorship would be acknowledged in the next issue. He also asked me if I had any more chess problems I could send him, and so in the next few months, I published several of my problems in the Manchester Guardian, which led to the publication of others in European and Canadian journals.

Some weeks later, a remarkable thing happened! The same problem — "Where is the White King?" — appeared in Martin Gardner's column in *The Scientific American* without mention of my name! It had been sent in by someone, with a note saying that he found the problem remarkable, but did not know who invented it. I knew nothing about this at the time, and probably would have never known, had it not been seen by the mathematician Mitch Taibleson, whom I had known as a student in my days at the University of Chicago. He wrote to tell Martin Gardner that the problem had been devised by Raymond Smullyan, who showed it to him in the days when he and Smullyan were students at the University of Chicago, and that this problem was only one of a large collection of unpublished problems.

If it had not been for that letter, my life may well have turned out quite differently! The letter led to a happy renewal of my friendship with Martin Gardner, who had already published several of my logic problems in his column. Martin urged me to stop dilly-dallying and get the book of my chess problems written! This goaded me to get to work.

I originally planned to incorporate all my retrograde problems into Arabian Nights stories, but then an expert on retrograde analysis named Mannis Charosch, who had seen some of my problems, sent me a paper he had written entitled "Detective at the Chessboard", which was an excellent introduction to retrograde analysis. The title intrigued me, and I thought, "Why not have an *actual* detective at the chessboard? And a perfect one would be Sherlock Holmes." I thus changed my plan and decided to divide my problems into two books, one on Sherlock Holmes, and the other on the Arabian Nights. The first I titled *The Chess Mysteries of Sherlock Holmes*, and the other I humorously titled *The Chess Mysteries of the Arabian Knights*. [Notice that I used "Knights" instead of "Nights".]

I then wrote to Martin of my plan and he then wrote to the chess editor of a well-known publishing company. The editor soon phoned me and was quite enthusiastic about the idea and asked me to send him the manuscripts when they were finished, which I did. Several weeks later, he informed me that although he was in favor of the books, the sales department turned them down! He explained that times had changed, and nowadays publishers wanted books with good commercial possibilities. And so here I was stuck with these two unpublished manuscripts! By then it was many years since I had devised my first retrograde chess puzzle and more than two decades since I had written my first retrograde chess puzzle book. I wondered if these books would ever see bookstore shelves!

Well, a few years later Fate fortunately intervened: As I have already mentioned, the logician, Professor Melvin Fitting, is a former student of mine. He studied with me when I was teaching at the Belfer Graduate School of Science in New York City in the mid 1960's, and after Mel obtained his doctoral degree there we became good friends and later collaborated on a book about set theory. Mel's father-in-law Oscar Collier was a literary agent and editor of Prentice Hall. He asked Melvin whether he would be interested in writing a book of logic puzzles. Mel told him that this was really not his line, but that Raymond Smullyan would probably be a good person for this. Oscar then phoned me and I told him about my chess books. He told me that he wasn't interested in chess problems, that what he wanted was logic problems. I thus went to work composing my first book of logic puzzles, which I entitled *What is the Name of this Book?*

After writing a few chapters, I took them to Oscar and brought the chapters of my chess books as well. As he had told me before, he was not interested in my chess books for Prentice Hall, but would propose my book of logic puzzles to them. As for the chess books, he, as a literary agent, would try to find another publisher.

He got Prentice Hall to publish my puzzle book, and sent my chess books to Alfred A. Knopf, where they were looked at by a senior editor named Ann Close. She quietly sent them back to Oscar Collier, telling him that chess was out of her line.

Then *What is the Name of this Book?* came out (in 1978) and received a rave review from Martin Gardner, who pronounced it "the most original, most profound, and most humorous collection of recreational logic and math problems ever written." Well, when Ann Close read this, she telephoned Oscar and asked for a second look at my chess books. Ann is an extremely clever lady who knew virtually nothing about chess at the time, but on her second look she applied herself so thoroughly that after a month she had enough expertise to not only understand my problems, but even to find mistakes in some of them!

I must tell you about a cute incident. Ann is from the South. Once when she was on vacation, I phoned her house in North Carolina, believing that she might be there. She wasn't, but I had a long conversation with her mother, to whom Ann was very devoted. At one point in the conversation, I asked her mother, "Are you as charming, beautiful, and talented as your lovely daughter?" She replied, "Oh, much more so!"

Knopf published both books, and that's the story of how my chess books got published. Incidentally, my special problem "Where is the White King?" appears on the front cover of the jacket of *The Arabian Knights* (finally published in 1981).

I subsequently published several more of my puzzle books with Alfred Knopf with Ann Close as the editor, one of the last of which is *The Riddle of Scheherazade and Other Puzzles, Ancient and Modern*. I later published more puzzle books, and others, on other topics, with other editors. As I may have already said, my puzzle books are more than just puzzle books; they are designed to lead the reader into deep results in mathematical logic through recreational puzzles. Let me give you some examples.

We return to my island of knights and knaves, where knights make only true statements, and knaves always lie and make false statements. Every inhabitant of the island is either a knight or a knave. The classical problem about this island is that some visitor went to the island and came across three inhabitants, A, B, and C. The visitor asked A whether he was a knight or a knave, but A answered so indistinctly that the visitor could not understand what A said. The visitor then asked B, "What did A say?" B replied, "A said that he is a knave." Then C said, "Don't believe B. He is lying!" The problem is, what is C? Is he a knight or a knave? Would the reader like to try figuring this out before reading further?

Well, here is the answer. No inhabitant of this island could say, "I am a knave," because a knight would never falsely claim to be a knave, and a knave would never truthfully claim to be a knave. Hence when B said that

A said that he was a knave, B was clearly lying. Thus C's statement that B was lying was true, and so C is a knight. We cannot tell the type of A, since we don't know what A really said.

I heard this problem many, many years ago, and years later it struck me that the island of knights and knaves would be an excellent setting to begin a course in mathematical logic. The logic of lying and truth-telling is one to which people easily relate, and so this would be a good way to begin the subject of propositional logic, based on the so-called propositional connectives — "not", "and", "or", "implies", and "if and only if".

A man named Larry visits the island and comes across natives A and B. Larry asks A, "Tell me something about yourselves." A replies, "I'm a knave and B is a knave." [Alternatively, he could have said, "Both of us are knaves."] What type is A (knight or knave) and what type is B?

On another occasion Larry came across two natives A and B, but this time A said, "At least one of us is a knave." [Alternatively he could have said, "Either I am a knave or B is a knave," which amounts to the same thing.] Which type is each?

On another occasion Larry came across two natives A and B, and A said: "If I am a knight, then B is a knave." Can it be determined what A is? What about B?

On still another occasion Larry came across two natives A and B, and A said, "B and I are of the same type, both knights or both knaves." [Alternatively, he could just as well have said "I am a knight if and only if B is a knight."] Can the type of A be determined? Can the type of B be determined?

Here are the answers. Consider first the case when A said, "I am a knave and B is a knave." Obviously no knight could say that, for he would then be claiming to be a knave. Thus A must be a knave. Since he is a knave, his statement is false — it is not really the case that both of them are knaves. Thus B is a knight. And so the answer is that A is a knave and B is a knight.

However, isn't there something wrong? We have already seen that no inhabitant can claim to be a knave — no knight would, and no knave would. How then can A say that he and B are both knaves? The fact is that the statement must be taken as a whole. If B is a knight, then it is false that both are knaves, i.e. the statement that "I am a knave and B is a knave" is simply false, even though part of the statement is true. [If I know French but don't know German, then I would be lying if I said that I know French and I know German, even though it is true that I know French.] The interesting thing is that although A can make the single statement, "I am a knave and B is a knave," he cannot make the two separate statements: (1) "I am a knave"; (2) "B is a knave". Thus conjunction has the curious property that if a constant truth-teller can assert the conjunction of two statements, then he can assert each of them separately, but for a constant liar, if he asserts a conjunction of two statements, it could be that he cannot assert each of them separately!

Now let us consider the second problem, in which A said, "At least one of us is a knave." If A were a knave, then it would be true that at least one of A, B was a knave, but knaves don't make true statements! Thus A must be a knight. Since A is a knight, then at least one of the two must be a knave, as A truthfully said. The knave must then be B. Thus the answer is the virtual opposite of the first one. In this case, A is a knight and B is a knave.

Now consider the third problem, in which A said, "If I am a knight, then B is a knave." This is a subtle and most interesting case. Suppose A is a knight. Then, like he truthfully said, if he is a knight, then B is a knave, and since A is a knight (by our supposition) then B is a knave. This does not prove that B is a knave; it only proves that *if* A is a knight [and lives on the island of knights and knaves, and has just said, "If I am a knight, then B is a knave"], then B is a knave. Thus we now know that if A is a knight then B is a knave. But since A said just that, then A spoke the truth, and so A is a knight! Since A is a knight, then as he truthfully said, if he is a knight, then B is a knave, and, since A actually is a knight, B must be a knave. Thus the answer is that A is a knight and B is a knave. [There is another way to see this solution. Is it possible for A to be a knave? Well, assume he is a knave. If A is a knave, what he said is false. And the only way for the implication "If I, A, am a knight, then B is a knave" to be false is if A is a knight and B is *not* a knave. But A can't be a knight under the assumption that A is a knave. Assuming A to be a knave has led us to a contradiction, so

A cannot be a knave and must be a knight. But if A is a knight he must speak truthfully and what he said implies that B must be a knave. Thus A is a knight and B is a knave.]

Lastly, what about the case when A claims to be the same type as B? Well, if B were a knave, A could never claim to be of the same type as B, as this would be paramount to claiming to be a knave. Hence B must be a knight. The type of A cannot be determined. He could be a knight who truthfully claims to be the same type as B, or he could be a knave, who falsely claimed to be the same type as B.

Here is another instructive problem. In a certain flower garden, each flower was either red, yellow, or blue. Also there was at least one of each of these three colors. One statistician observed that whichever three flowers were picked, one was bound to be red. Another statistician observed that whichever three flowers were picked, at least one was bound to be blue. Does it logically follow that whichever three flowers were picked, one was bound to be yellow? Yes, it does! Why?

Before giving the answer, let me tell you the story of the statistician who told a friend that he never took planes. When asked why, he explained that although the probability that there would be a bomb on a plane was low, it was still too high for his comfort. Well, a couple of weeks later, the friend met the statistician on a plane and asked him why he had changed his theory. The statistician replied, "I subsequently computed the probability that there would simultaneously be two bombs on the plane. The probability is low enough for my comfort, and so I now carry my own bomb!" [Incidentally, this is a typical freshman fallacy!]

Now for the answer to the flower problem. The answer is that if you pick three flowers, one of them must be yellow, because there are only three flowers in the entire garden! Here is why: Suppose there were more than three. Then at least two of them would have to be of the same color (since there are only three colors). Suppose there are two reds. Then one could pick these two together with a yellow, thus avoiding a blue, which is contrary to the condition that of any three flowers, one has to be blue. Thus there could not be two reds. Suppose there were two blues. Then one could pick them with one yellow, thus avoiding a red, which is again contrary to the given conditions. If there were two yellows, one could pick them together with a red, thus avoiding a blue (or alternatively, one could pick them with one red, avoiding a blue). Thus there are only three flowers in the garden.

Here are two more problems. In a certain convention of 100 scientists, each one was either a chemist or a physicist, and none of the scientists was both a chemist and physicist. It so happens that given any two of the scientists, at least one was bound to be a physicist. How many were physicists and how many were chemists?

Here is the answer. For the problem of the 100 scientists, some people have given the wrong answer that there were 50 of each. This can't be right, for if there were fifty chemists, one could pick two of them and avoid a physicist, contrary to the given fact that for any two, one must be a physicist. Indeed, to say that given any two, at least one is a physicist is tantamount to saying that no two are chemists! Thus there can be only one chemist, and the other 99 are physicists.

I love Thurber's characterization of logic in his story, "The Thirteen Clocks", in which he says, "Since it is possible to touch a clock without stopping it, it follows that one can start a clock without touching it. That is logic, as I understand it." I also like Tweedledum's characteristic of logic in *Alice's Adventures through the Looking Glass* of Lewis Carroll: "If it was so, it might be, and if it were so, it would be, but since it isn't, it ain't. That's logic."

I like even better the definition of logic given by Ambrose Bierce in the book *The Devil's Dictionary*: "Logic is the art of thinking and reasoning in strict accordance with the limitations and incapacities of the human misunderstanding. The basis of logic is the syllogism, consisting of a major and a minor premise and a conclusion thus:

Major Premise: Sixty men can do a piece of work sixty times as quickly as one man.
Minor Premise: One man can dig a post-hole in sixty seconds.
Therefore,
Conclusion: Sixty men can dig a post-hole in one second."

Speaking of syllogisms, it is important to realize the difference between a syllogism being *valid*, and a syllogism being *sound*. A syllogism is called *valid* if the conclusion really does follow logically from the premises, regardless of whether the premises are true or not. A syllogism is called *sound* if it is valid, and when the premises are also true. The typical sound syllogism is:

> All men are mortal.
> <u>Socrates is a man.</u>
> Therefore, Socrates is mortal.

The following syllogism, which is clearly not sound, is nevertheless valid:

> All bats can fly.
> <u>Socrates is a bat.</u>
> Therefore, Socrates can fly.

This syllogism is obviously not sound, since it is not true that Socrates is a bat, but it is valid, since the conclusion is a logical consequence of the premises — that is, if Socrates actually were a bat, then he would be able to fly.

I love syllogisms which on the surface appear to be invalid, but which in reality are valid. Here are two examples:

(1) Everyone loves my baby.
   <u>My baby loves only me.</u>
   Therefore, I am my own baby.

This sounds like a silly joke, doesn't it? However, it really is valid. Since everyone loves my baby, then my baby, being a person, loves my baby. But since my baby loves only me, I am the *only* person my baby loves, i.e. anyone my baby loves must be me. Since my baby loves my baby, then the one and only one person my baby loves must both be me *and* my baby. Therefore my baby and I must be the same person!

(2) Everyone loves a lover.
   <u>Romeo loves Juliet.</u>
   Therefore Iago loves Othello.

Why is this valid? Well, since Romeo loves Juliet, then Romeo is a lover (by the definition of "lover" above). Since Romeo is a lover, then everyone loves Romeo. Hence everyone is a lover. Hence everyone, being a lover, is loved by everyone. Thus everyone loves everyone, and so, in particular, Iago loves Othello.

Here is a cute syllogistic joke:

> Some cars rattle.
> <u>My car is some car.</u>
> So no wonder my car rattles!

Someone once asked the philosopher and logician Bertrand Russell, "What is new in the conclusion of a syllogism? Isn't the information already implicit in the premises?" Russell answered that although the conclusion might not have any *logical* novelty, it certainly can have *psychological* novelty, and he told the following story to illustrate the point:

At a certain party, one of the guests told a risqué story. Someone else told him, "Be careful! You forgot that the Abbot is here!" The Abbot then said, "We men of the cloth are not as naïve as some of you think. Why, my very first penitent was a murderer!" Shortly afterwards, an aristocrat arrived late at the party. Someone wanted to introduce him to the Abbot and asked him if he knew the Abbot. The aristocrat replied, "Of course I know him. I was his first penitent!"

Let us recall the lovely incident in Plato's dialogues in which the Sophist gives the following proof to Socrates that he was the son of a dog!

> *Sophist*: Do you have a dog?
> *Socrates*: Yes.
> *Sophist*: Male or female?
> *Socrates*: Male.
> *Sophist*: Does he have any puppies?
> *Socrates*: Yes, I saw him and a female dog come together.
> *Sophist*: Then he is a father.
> *Socrates*: Yes.
> *Sophist*: And he is yours.

Socrates: Yes.

Sophist: Therefore he is your father.

[I am surprised that this could work in English!]

Once Socrates chided the sophist Protagoras for taking money for teaching wisdom. Protagoras then told Socrates that if the student was not satisfied with the teaching, his money would be refunded in full. Well, when I read this, I thought of the following scenario: A student goes to Protagoras, and at the end of his studies complains that he did not learn enough and asks for his money back. Protagoras asks him whether he could give a good argument why he should get his money back. The student gives an excellent argument, upon which Protagoras says, "You see the dialectical skill I have taught you!"

Protagoras also taught law, and there is an interesting paradox known as the *Protagoras Paradox*. A student comes to Protagoras, but has no money to pay. Protagoras recognizes that the student is quite talented, and makes the following agreement with him. The student does not have to pay for the course of instruction, but at the conclusion, when the student is then qualified to practice law, after winning his first case he will pay Protagoras a specified amount. Well, after the conclusion of the instruction, the student didn't take any cases, and so Protagoras sued him. The student acted as his own lawyer for the defense. The student argued, "Either I win this case, or I lose this case. If I win, then I don't have to pay, since this is what the case is about. On the other hand, if I lose this case, I will not yet have won my first case, and so I then don't have to pay. Thus, regardless of whether I win or lose, I don't have to pay!" Protagoras then said, "He has it all wrong! If he loses the case, that means he has to pay me, for that is what the case is all about. On the other hand, if he wins the case, he will have won his first case, and so he must pay me. In either case, he must pay me!" How should this case be resolved? The best answer I ever got was from a lawyer. He said that the student should win the case and not have to pay. Now that he has won the case, Protagoras should sue him a second time!

Speaking of lawyers, do you know why it is safe for a lawyer to swim in shark-infested waters? Professional courtesy!

I also like the story of the lawyer who said to his client, "You can ask me questions and I will answer. I charge one hundred dollars per question." The client replied, "Isn't that rather excessive?" The lawyer said, "No! Now, what's your second question?" The client said, "Hey, you're not going to charge me for that question, are you?" The lawyer said, "Yes! Now what's your third question?"

* * *

Before returning to my Princeton days, let me tell you that sometime before I went to Princeton, there was a little girl there who was doing badly in mathematics. In the space of a few months, she was doing much better. When her mother asked her why she was improving so much, she said, "I heard there was a teacher here who teaches real good! I go to his house after school and he helps me. He really teaches real good! I forget his name. It is Ein something … something like Einstock." It was indeed Albert Einstein! She went to his house frequently, and of course he helped her.

One incident I heard about Einstein really delighted me! He told a colleague that he didn't like teaching at a co-ed school. When asked why, he explained that with all the pretty girls in the room, the boys wouldn't pay attention to mathematics and physics. The colleague said, "Oh, come on Albert, surely all the boys would listen to what *you* have to say!" Einstein replied, "Oh, such boys are not worth teaching!"

7 | After Graduation

After getting my Ph.D., I taught for two years at Princeton. I subsequently taught at New York University, the University of New Paltz, Smith College, the Belfer Graduate School of Science, Lehman College, the CUNY Graduate Center, and Indiana University Bloomington. At Lehman College I taught undergraduate courses. My method of giving final grades was perhaps a bit unusual. I would ask the student what grade he or she believed they deserved. If that grade was the same as the one I thought they deserved, that was it. If the student's evaluation was lower than mine, I would give the student my grade. If the student's evaluation was higher than mine, we would then argue. The interesting thing is that in an overwhelming majority of cases, our evaluations were the same! I recall one particular case in which the student (whom I suspect was psychotic) believed she should get an A, whereas she was clearly an F student. I gave her a C.

At the CUNY Graduate Center, where of course I taught only graduate courses, I repeatedly got into trouble with many faculty members because I was too easy a grader! Yes, I was indeed an easy grader. I have always regarded the grading system as unwholesome. For a teacher to report a bad grade of a student seems to be like a doctor breaking the Hippocratic Oath! Yes, I can understand an institution or business giving an *incoming* examination for an applicant, but for a *teacher* of a student to report his or her grade seems to me to put the teacher in an adversarial relation to the student. Since I could not change the grading system, I thought that the best I could do was to grade as liberally as possible. I guess I am like the late eminent philosopher Alfred North Whitehead, who almost always gave A's to his students. As an experiment, one student turned in a lot of nonsense, just to see what would happen. Whitehead gave him a B!

As I said, I got into great trouble at the Graduate Center because of my grading policy. It really raised a lot of turmoil. Fortunately I received a most attractive offer from Indiana University for a distinguished rank professorship. I then happily left CUNY for Indiana University in Bloomington, Indiana, where I remained for several years. I particularly enjoyed Bloomington for its psychologically sunny climate!

I must tell you one particularly memorable incident. At Indiana I taught both graduate and undergraduate courses. Among my undergraduate courses was a course in logic for liberal arts students. I used my own *What is the Name of this Book*? as a starter, followed by some propositional and first-order logic, and included some topics on infinity. On my final exam, in order to give those students who were poor in mathematics and logic an opportunity to raise their grade, my final question of a group of four was "Write on anything you want." Well, one student wrote a story so excellent that had he been an F student, I would have given him an A, just for that story. As a matter of fact, he was also an A student. Here is the story he wrote:

> Once upon a time there were two tribes, the altruistic tribe and the selfish tribe. The altruists didn't want to do anything for themselves. They wanted only to do things for Society. The selfish ones not only didn't want to do anything for Society; they had a positive *aversion* to doing anything for Society. One day the two tribes got into a war, and everybody was killed, except for one person on each side. They both pointed their guns at each other, and the altruist thought, "If I shoot him, I will be the whole of Society, and anything I do for Society I'll be doing for myself, and I don't want to be selfish!" The selfish one thought, "If I shoot him, I'll be the whole of Society, and anything I do for myself, I'll be doing for Society, and I don't want to do anything for Society!" And so they didn't shoot each other!

That was the story. I wish I could remember the name of the student. I really think the story is publishable. If the student who wrote the story ever reads this, I hope he will get in touch with me.

Once when I was teaching a graduate course in mathematical logic, a particular young woman came in late and asked, "Professor Smullyan, can I please have the notes?" I replied, "I'll give them to you if you are good!" She, being a very bright girl, asked, "What does it mean to be good?" I replied, "It means not knowing what it means to be good!" [The students really enjoyed that one.]

Indiana University has a wonderful music department, equal to our best conservatories. The famous violinist Joshua Bell was a student there, and his teacher was the eminent Joseph Gingold, whom I had the pleasure of meeting several times. Many of the music students were superior to many of the faculty members! I recall that a chamber music group of faculty members once performed a certain work, and a few days later, the same work was performed by a group of students, only so much better! Every day was filled with several concerts, and I recall that I once heard three concerts in one day.

While on the subject of music, let me tell you some of my favorite music anecdotes!

Once Beethoven was giving a lesson to a student who was playing one of Beethoven's sonatas. When the student finished, Beethoven said, "The way you played it was not at all what I intended when I composed it, but I like your interpretation at least as much as what I had in mind!"

This bears an important similarity to an incident in which Chopin was playing his Barcarolle for a group of friends. Towards the end of the piece, the score is written *ff* (*fortissimo* — very loud), but Chopin played the section very softly. When asked why he had deviated so much from what he had written, he replied, "Oh, it could be played either dramatically loud or dramatically soft. Either way works. I happen to be better at playing *pianissimo* than *fortissimo*, and so that's the reason I played it that way."

These two incidents so beautifully illustrate the fact that composers are so much more tolerant of departures from their written scores than are music critics, music teachers and many performers! In the Chopin case, if any pianist today would play that passage soft, instead of loud as it is written, he or she would get clobbered by the music critics!

Coming back to Beethoven, many of you know the story of how Beethoven once got into an argument with a prince. At one point the prince drew himself up stiffly and said, "Do you realize that I am a prince?" Beethoven replied, "There are many princes, but there is only one Beethoven!" I like it when people honestly express pride in what they do!

I love the story of Brahms and an amateur string quartet. Brahms knew four chamber music players. They were not very good musicians, but they were such nice people that Brahms liked to associate with them. They once decided to surprise Brahms, and so they spent six months assiduously practicing his latest string quartet. One night at a party they cornered him and said, "Johannes, come upstairs. We have a surprise for you!" They went upstairs and the four started playing the quartet. The first movement was as much as poor Brahms could bear! He politely smiled and walked out of the room. The first violinist ran after him and asked, "Johannes, how did we do? Was the tempo alright?" Brahms replied, "Your tempos were all good. I liked yours the best!"

On another occasion Brahms was leaving a party, and just before he left, he said to the group, "If I have neglected to insult anybody, I hope he will forgive me the oversight!"

Franz Schubert would often forget things that he had previously composed. There is the story told that about five years after having written a certain composition, someone played it for him at a gathering, and Schubert then said, "That's not bad. Who wrote it?"

There is another, less-known, aspect of Schubert that is quite different from the sweet, kind, loving, lyrical Schubert that we all love so well! One late night he and some friends were in a tavern, and they — particularly Schubert — were drinking heavily. At one point, some professional artists, members of the Opera House orchestra, came in and caught sight of Schubert, and went over to him and smothered him with flattery. It turned out that they were very anxious to have a new composition for their concert, with solo passages for their particular instruments. And they were sure that Schubert would comply. But Schubert turned out to be anything but accommodating! After many entreaties, Schubert said, "No! For you I will write nothing!"

"Nothing for us?" they asked.

"No, definitely not!"

"And why not, Herr Schubert?" came the reply. "I think we are just as much artists as you. No better artists can be found in the whole of Vienna."

"Artists!" cried Schubert. "Artists?" he repeated. "Musical hacks are what you are! Nothing else! One of you bites the brass mouthpiece of a wooden stick, and the other blows out his cheeks on the horn! You call that art? It's nothing more than a trade that earns money! You call yourselves artists? Blowers and fiddlers are what you are, nothing more! I am an artist. I! I am Franz Schubert, whom everybody knows and respects! One who has written great things and beautiful things that you are not capable of understanding, and one who will continue to write great things — cantatas, quartets, operas and symphonies. I am Schubert! Franz Schubert! And don't forget it! If the word *art* is mentioned, it is *I* they are talking about, not you worms and insects, who demand solos for yourselves that I shall never write for you! You crawling, gnawing worms that ought to be crushed under my feet — the foot of a man who is reaching to the stars. Yes, to the stars, while you poor puffing worms wriggle in the dust, and with the dust are scattered like rats!" When I read this, I was of course most surprised and shocked, but I was more amused than shocked. I understand that the next day, when Schubert sobered up, he did write some music for the group.

I must diverge for a moment and tell you a cute incident from the present. I have a friend, Lisa Kovalik, a pianist who teaches at the Juilliard School. Once she was at my house and I made a recording of her playing the first movement of the Beethoven Opus 109 Sonata. Some weeks after that, I brought the CD I had made to a gathering of friends at her house. I was in a mischievous mood at the time, and thought I would tease her a little. So I showed them all the CD, and before playing it, I said, "Here is a recording I made of me playing a movement of a Beethoven sonata, and I play it much better than Lisa!" Lisa got quite upset over this, and could not get over my colossal egotism! After playing it, to her relief, she realized the joke that it was really her recording.

Incidentally, I, and a mutual friend, once wrote the following little verse about Lisa:

Lisa,
We like to please her,
We also tease her,
We squeeze her,
And don't release her!

Coming back to musical anecdotes, some music critic wrote an extremely bad review about Franz Liszt, telling how bad he was, both as a pianist and as a composer. Several friends of Liszt urged him to publish a rebuttal. Liszt finally agreed to do so, and here is what he wrote: "I fully agree with everything you said, and I am happy that I have been helpful in your making a name for yourself."

Liszt was a very gracious person, and also had a good sense of humor, as the following incident will reveal. He and Clara Schumann were close friends, despite the fact that Clara had great contempt for Liszt's musical compositions, and Liszt knew it. [However, she had great respect for his piano playing, especially his remarkable technique.] Once Clara had to give a concert and Liszt escorted her to the concert hall. When they looked in, they saw and heard that the audience was extremely rowdy. Clara Schumann said, "What I have prepared is much too good for this crowd!" Liszt ironically responded, "Then why don't you play some bad stuff by Liszt?"

I recently learned some facts about Paganini that increased my respect for him enormously! He was really far more than a mere showman. He liked to play chamber music, particularly Beethoven quartets. There was a certain singer at the time who had a good voice, a good technique, but was not at all musical. The way that Paganini described her lack of musicality was most interesting! He wrote, "I attended her concert and was bored. She had a good voice and a good technique, but she is lacking in musical philosophy!"

There was a 19th century singer whose voice was also good but not very musical. The music critic Ernest Newman described her voice as "uninterestingly perfect, and perfectly uninteresting!"

Another singer (Nellie Melba) was a good singer but a hard-nosed business woman. Her manager once told her that the salary she was asking for was more than that of the President of the United States! Upon which she replied, "OK, then get the President to sing for you."

When the composer Meyerbeer died, one of his admirers wrote an elegy for him and showed it to Rossini and asked his honest opinion of it. After looking at the score, Rossini said, "To be perfectly honest, I think it would have been better if you had died and he had written the elegy."

In a performance of an opera by Massenet, a certain tenor had sung flat throughout. After the performance, the tenor came to Massenet, hoping to receive congratulations, and said, "I hope you were pleased with my performance!" Massenet replied, "Delighted, delighted! But how could you sing with that dreadful orchestra who accompanied you half a tone sharp all evening?"

Now for Richard Wagner! Of his music, Mark Twain said, "It's probably not as bad as it sounds!"

There is the story told that Wagner was once walking down a street in Berlin and came across an organ grinder grinding out the overture to Tannhauser. Wagner told him, "As a matter of fact, you're playing it a bit too fast!" The organ grinder tipped his hat and said, "Oh, thank you, Herr Wagner! Thank You!" The next day, Wagner came across him again, this time playing the overture at the correct tempo, with a sign: PUPIL OF RICHARD WAGNER.

One biographer wrote, "The marriage of Wagner and Cosima was a happy one. They had a common interest: Both adored Richard Wagner."

I once saw a play in which the entire action took place in Hell. A visitor was very surprised when told by the Devil that Wagner was in Hell. "Of course," said the Devil. "He was a vicious anti-Semite!" The visitor said, "But he wrote such beautiful music!" "Oh," said the Devil, "his music went to Heaven. *He* went to Hell!"

Here is a very funny story about the composer Anton Bruckner. He was so naïve that he once went to an orchestra concert and liked the conductor so much that he later went backstage and gave him a tip! The conductor was so moved that he later drilled a hole in the coin and wore it around his neck the rest of his life!

Next, I must tell you a very moving story about Puccini (the composer of *Madam Butterfly*). One day he heard a man's voice singing outside his window. It expressed such beauty and feeling that Puccini was moved to tears. When he looked out the window, he saw that the man was a convict who had been ordered to mend the roads. His feet were in chains. Puccini could not believe that a man who could sing with such soul could really be a criminal. Puccini interceded with the Queen of Italy and obtained a pardon. The condemned man was permitted to return to his wife and child, and forever after revered Puccini with dog-like devotion.

The conductor Sir Thomas Beecham had a fabulous memory. When he once memorized an opera score, he never had to look at it again. Years later he could conduct it from memory. He was also incredibly absent-minded. Once he was about to conduct an opera, and with his baton in his hand, he asked the concert master, "By the way, what opera are we doing tonight?" He also had an acidic sense of humor. One of his lady opera singers once introduced him to her twelve-year-old son. Beecham asked the boy whether he could sing. When the boy said he couldn't, Beecham said, "Oh, I see it runs in the family!"

Speaking of sarcastic humor, a true master of that was the pianist Leopold Godowsky. One night a composer friend of his phoned him and told him that he must come over and hear his last composition. "Your very last?" asked Godowsky, "Yes," said the friend. "Good!" said Godowsky.

On another occasion, Godowsky visited another composer friend, and when he arrived at the house, he found his friend composing merrily away, surrounded by many symphonic and opera scores. Godowsky said, "Oh, I thought you composed from memory!"

The delightfully eccentric pianist Vladimir de Pachmann was once giving a concert when he spied Godowsky in the audience. He stopped playing, and said, "I won't play any more until Godowsky comes up on the stage!" Embarrassed, Godowsky went up and de Pachmann planted a kiss on his forehead and said, "Godowsky, the greatest pianist in the world! But de Pachmann plays more beautifully!" [Which was true, by the way.]

Turning to jazz, I once heard a seminar of jazz players, each of whom was giving his definition of jazz. The one I liked best was that of a black jazz musician who said, "To me, jazz is a feeling. It's the feeling you get when you know that you're going to get a feeling."

The great 19th century mathematician Felix Klein was once at a party in which people were discussing the relation between music and mathematics, both interests and aptitudes. Klein grew more and more puzzled and finally said, "I don't understand, gentlemen. Mathematics is beautiful!"

Going from mathematicians to physicists, the eminent physicist Paul Dirac once gave a lecture, at the end of which he asked whether there were any questions. One student raised his hand and said, "Professor Dirac, I didn't understand your proof of Theorem 2." Dirac answered, "That's not a question."

The physicist Wolfgang Pauli was a terror at lectures, because he always found errors in the speakers' proofs. The story was told that when Pauli died, he went to the pearly gates and Saint Peter said to him, "Of course you are welcome here, but there will be a problem of your adjustment! I don't see that there is anything here that you can do, since the answers to all possible questions are already known. Anything you want to know, you simply ask the Master, and He will simply tell you. You had better meet the Master now." So saying, Saint Peter led Pauli into a library and there was God seated behind a desk in front of a blackboard. God then said, "It's like Peter said! There will be a problem of your adjustment here, since whatever you want to know, I have the answer. Is there anything you want to know?" Pauli replied, "Well, Lord, you know that all my life I have been trying to find the value of the fundamental constant. What is it, and why?" The Lord smiled and said, "Well, when I created the universe, I didn't want the value of the constant to be that obvious, yet it is not really that difficult to figure out." God then went to the blackboard and wrote down a few equations, and Pauli nodded. Then God wrote down some more equations, and concluded, "Therefore this is the value of the constant." Pauli thought for a while and finally said, "That proof is very ingenious! But it's wrong!"

8 | Diagonalization, Love, Logic, a Letter

This chapter will involve stories from my life — on the topics mentioned in the chapter title — which will have occurred in many places and times. In it I will frequently not mention when or where the thoughts or events took place, and may not even relate them in the order the events took place. Realize that most of the thoughts of mathematicians and scientists are often seen by them as not lying in any place or time!

Let me now tell you how I, together with Dr. Bruce Horowitz, a former student of mine, started a group called the *Diagonalization Society*. It came about this way. There is a term *diagonalize* in mathematical logic, which is a device for achieving self-reference, and which played a key role in the proof of Gödel's incompleteness theorem. Well, once, when Bruce was my student, he said, "Professor Smullyan, I think you like to diagonalize people!" Just what is meant by "diagonalize" in this context is hard to explain. It is a sort of composite of *trick, fool, confuse, confound, mystify, get the better of,* and other things. Regardless of what it exactly means it is a term that Bruce and I have been using for years, and we constantly try to diagonalize each other! A few months ago, it occurred to us to form a group called *The Diagonalization Society*, which we posted on the Internet. We have had several contributors. I will now give you the description there and then several of the entries.

DIAGONALIZATION AND OTHER LOGICAL MATTERS
BY THE DIAGONALIZATION SOCIETY

FOUNDER: Bruce Horowitz
EDITOR-IN-CHIEF: Raymond Smullyan
PURPOSE OF THE SOCIETY: To diagonalize.
OUR MOTTO: Be wise! Diagonalize!
EDITORIAL NOTE: Much to the annoyance of some of the other members, the editor has made some minor modifications to some of the contributions.
EDITORIAL NOTE 2: An editor is best defined as "One whose function is to ruin an author's manuscript."

Now let the fun begin!

There was a composer named Arnie,
Who used to work in a carny.
He diagonalized Ray,
One remarkable day,
And made him think he was blarney.

There was a man named Newton,
Not to be confused with Rasputin,
When the apple hit betwixt his eyes,
He cried, "Eureka! I'll diagonalize!"
And with that there's no disputin'!!!

George Boole
Was nobody's fool.
When diagonalizing at night,
He chose only the bright,
And whisked them off to school.

When diagonalizing by day,
He never chose Ray,
Because he would say,
"That awful Ray!
That incorrigible Ray!
That Goddamn Ray!
He will never obey!

He refuses to play,
He refuses to play by the Rule!"

There was a fellow named Brian,
Who always kept tryan and tryan,
To diagonalize Ray,
Just about every day,
Until poor Ray was cryan!

Yes, the fellow named Brian
Really had the force of a lion.
Day by day,
He would diagonalize away,
Until all the animals were dyan.

Another thing about Brian,
He never could be caught lyan.
Everything he said was the truth.
Yes, in sooth,
He was really loyal to Zion.

There was a logician named "Gödel",
Whose name was pronounced like "Girdle".
His first name was "Kurt".
He was very alert,
And could overcome any logical hurdle.

There was a logician named "Ruth"
Who had a diagonal tooth,
Which could easily diagonalize
All manner of lies,
And make them seem like the truth.

As for the logician Rice,
His theorems were really quite nice.
He carried curry
With Haskell Curry
And always took his advice.

In fact the two, one night,
Got together and began to write

A joint paper, whose theorems were right.
And so, to my great delight,
A paper by Curry and Rice!

There was a logician named Ray,
Who thought diagonalization all day.
They tried to enumerate
The ways he could ruminate,
But Ray always found a new way.

There was a logician Raymond Merrill.
You diagonalized him at your peril.
If you have the nerve
To throw him a curve,
He'll render your arguments sterile.

Raymond was seldom outsmarted.
He was wise-headed and youthfully-hearted.
He founded a nation
Based on diagonalization,
But when you get there, you're back where you started.

There was a logician named Ray,
Who diagonalized in Yiddish one day.
The Torah, Ray beams,
Is not what it seems,
And the Rabbi shouted "Oy vey!"

The versatile Bertrand Russell
Enjoyed any sort of a tussle.
He could argue with force
On math or divorce,
And he sexually had plenty of muscle!

There was a featherless biped,
Whose name was Alfred Whitehead.
He came home from a ball,
And began to scrawl,
"I'm drunk and have a light head."

The illustrious Dana Scott
Used to diagonalize a lot.
He would diagonalize all his friends,
For God know what ends,
Until they all really got hot!

> As for the logician Mike Dunn.
> He was constantly on the run.
> And would diagonalize Ray
> Day by day.
> And together they had lots of fun.

There was this fellow named Pete,[2]
Who very much wanted to eat.
But a fly came by
And sat on his pie.
Which he unfortunately was not able to beat.

Now this man named Pete,
Always dressed very neat.
He would diagonalize at night,
Even the most bright.
A truly remarkable feat!

As for his wife named Dot,[3]
She could diagonalize any kind of a knot,
And make it think
It could turn pink,
And then become red hot.

> The Puzzle Lady Kate
> Simply could not wait
> To diagonalize Ray,
> Which made him say,
> "Diagonalization is my fate!"

---

[2] Peter Denning
[3] Dorothy

Now, Ray's students:

Bruce Horowitz

There was a logician named Bruce,
Who would diagonalize on the loose.
Then one fateful day
Tried to diagonalize Ray,
Who said, "Bruce, I see through your ruse!"

There was a logician named "Bruce"
Who would always think, "What's the use?"
Of diagonalizing Ray, because Ray would say,
"You obviously have a screw loose!"

Melvin Fitting

There was a logician named Mel,
Who thought diagonalizing was swell.
To all those unwitting
He thought it was Fitting
To fool them before they could tell.

There was a logician named Fitting,
Who once proved a theorem while sitting.
He jumped to his feet,
And ran down the street,
Yelling, "Eureka," before the ground hitting.

A logician's wife named Roma,
Admired the writings of Homer.
She thought it was Fitting
Not to always do knitting,
But to sometimes allow Melvin to comb her.

Malgosia Askanas

There was a logician Malgosia,
Who diagonalized in Nova Scotia.
Then during one gig
Diagonalized a pig
And made it think it was kosher!

Robert Cowen

> There was a logician named Bob,
> Who diagonalized while on the job.
> His students would yearn
> That someday they'd learn
> How their senses he'd routinely rob.

Fred Halpern

> There was a logician named Fred,
> Who diagonalized everyone's head.
> From reports, it seems
> He diagonalized their dreams
> And caused them to fall out of bed.

Sue Toledo

> There was a logician named Sue,
> Who diagonalized whoever she knew.
> She was not satisfied
> Until they all cried,
> " 'Tis false if and only if true!"

Next, here is a letter I wrote to Bruce:

Dear Bruce,

I will prove something that should surely be of interest to you! First for some definitions: By a META-DIAGONALIZER I mean one who has diagonalized at least one diagonalizer. By an AUTO-DIAGONALIZER I mean one who has diagonalized himself. By a META-AUTO-DIAGONALIZER I mean one who has diagonalized at least one Auto-Diagonalizer.

I will now prove to you that I am not only a diagonalizer, which you already know, but am in fact a Meta-Auto-Diagonalizer! Indeed, I once diagonalized God into diagonalizing himself! It happened this way.

To begin with, it is obvious that if there is no God, then the physical universe came into existence all by itself — it was not created by any God. I am assuming here, as did Thomas Aquinas, that God exists and that He can do anything that is logically possible. [As Aquinas said, it is blasphemous to believe it a limitation of God's power that He cannot do things that are logically impossible.] Now could God see to it that a universe would come into existence which was not created by God? Strange as this would be, there is no reason to believe that it is logically impossible, hence God could do that if He wanted to.

Now, at this point I must unfortunately differ from the Biblical account of the origin of the universe. The real truth of the matter is this: Before there was any physical universe, there was just God and spiritual beings. I was one of them, and was just as mischievous then as I am now! I once said to God, "Is it really true that You can do anything that is logically possible?" He replied, "Of course!" I then said, "OK, I dare You to see to it that a universe comes into existence all by itself — one that was not created by You or any other being!" To my surprise He did just that! But I didn't like that universe and begged Him to cancel it, which He kindly did. I then said, "Very good, but even better, I would like to see whether You could see to it that a universe comes into existence that was not only not created by You or any other being, but is the very universe that You would have created, had you created one!" Incredible as it might seem, He did just that! But surely that was an act of diagonalization, and so I had diagonalized him into diagonalizing himself! Q.E.D.

Postscript. The second universe I told you about, which came into existence without having been created by God or any other being, is the very one which we are in! Thus the atheists are absolutely right in that our universe was never created — it came into existence all by itself! But the amazing thing is that if God had created a universe, it would have been this one! Thus our universe, though not created by God, behaves just as it would had God created it! Isn't that remarkable?

Can anyone prove that this is not the case?

A Gothic Quiz

(1) What Gothic people were very active?
(2) What Gothic people had magic powers?
(3) What Gothic people suffered from vertigo?

 (4) What Gothic people were effervescent?
 (5) What Gothic people were very macho?
 (6) What Gothic people were constantly urinating?
 (7) What Gothic people were very feminine?
 (8) What Gothic people were rich and aristocratic?
 (9) What Gothic people never hit their target?
(10) What Gothic people were very cowardly?
(11) What Gothic people were quite gruesome?
(12) What Gothic people were often in a fog?
(13) What Gothic people were always striking people?
(14) What Gothic people were very gaudy?
(15) What Gothic people suffered from epilepsy?
(16) What Gothic people were named after the nickname of a British Queen?
(17) What Gothic people wore boxing gloves?
(18) What Gothic people had problems with sand?
(19) What Gothic people were absent?
(20) What Gothic people were always giving examinations?

Answers

 (1) The Busy Goths
 (2) The Wizzy Goths
 (3) The Dizzy Goths
 (4) The Fizzy Goths
 (5) The His-y Goths
 (6) The Pissy Goths
 (7) The Missy Goths
 (8) The Ritzy Goths
 (9) The Miss-y Goths
(10) The Sissy Goths
(11) The Grizzly Goths
(12) The Misty Goths
(13) The Hit-sy Goths
(14) The Glitzy Goths
(15) The Fitsy Goths

(16) The Lizzy Goths (Queen Elizabeth)
(17) The Fisty Goths
(18) The Gritsy Goths
(19) The Missing Goths
(20) The Quiz-y Goths

From Arnold Vance

(1) Who was the most enlightened Goth?
(2) Where did Supergoth live?
(3) What do Goths fly on?
(4) What is a young Goth called?
(5) Who gathers Goths together?
(6) What's a Gothic curse?
(7) What did God say to the Goths?
(8) What did Bach compose for the Goths?

Answers

(1) Gothama
(2) Gotham
(3) Gothamer wings
(4) A Gothling
(5) A Gothherd
(6) Goth-dammit!
(7) Goth forth and multiply
(8) The Gothberg Variations [This is my favorite!]

My Reply to Arnold

(1) What does a wealthy Goth say?
(2) What camphor materials do Goths put in their closets?
(3) Where do Goths go when they are sick?
(4) With what do Goths hit balls on the ground in a famous game?
(5) What do Goths call a pleasant occasion when they get together?
(6) What opera did Wagner dedicate to the Goths?
(7) What place do holy Goths visit?
(8) Why should you believe all this?

Answers

(1) I Goth lots of money!
(2) Goth balls
(3) The Gothpital
(4) Goth clubs
(5) A good Gothering
(6) The Gotherdammerung
(7) Gothsemane
(8) Because it's the Gothpel truth!

From Malgosia Askanas

(1) Where do Goths get their coffee from?
(2) What field of endeavor is dedicated to Gothic food preparation?
(3) What Spanish city is particularly hospitable to Goths?
(4) What European state is particularly hospitable to Goths?

Answers

(1) Gothta Rica
(2) Gothtronomy
(3) Taragotha
(4) Gothnia

From Bruce Horowitz

Here are a few more for Goths, some for Huns, and one for both:

(1) What does a Goth say when frustrated?
(2) Where does a Goth go for sex?
(3) Where was the center of the Goth mathematical world 100 years ago?
(4) What does a Hun call his wife?
(5) How many cents in a dollar?
(6) What is the fear of Huns called?
(7) What military vehicles do Huns drive?
(8) What were the methods that Huns and Goths used to obtain food?
(9) What make of car do Huns use for a hearse?

Answers

(1) Goth-darn it!
(2) To a Gothel
(3) Gothingen
(4) Huney
(5) One Hundred
(6) Hundread
(7) Hunvees
(8) Hunting and Gothering
(9) A Hundie

To Which I Might Add

(1) Why did the Huns eat so much?
(2) What was the Huns' favorite food?
(3) What form of execution did the Huns particularly dread?
(4) What country particularly welcomed the Huns?
(5) What opera was written for the Huns, and by whom?
(6) Who wrote fairy tales for the Huns?
(7) Why did the Huns never cheat?

Answers

(1) Because they were Hun-gry
(2) Hunburgers
(3) Being hung
(4) Hun-gary
(5) Hunsel and Gretel by Hunperdick
(6) Huns Christian Anderson
(7) Because they believed that Hunesty is the best policy

* * *

Now turning to the topic of love, and coming back again to my days as a graduate student at Princeton, I would frequently visit New York City. On one of my visits, I met a charming lady musician. On my first date with her, I asked her, "I'm going to make a statement. If the statement is true, would you give me your autograph?" She replied, "I don't see why not." I asked, "If the statement is false, then you don't give me your autograph, OK?" She assented. And so, I told her, "Remember, a true statement gets an autograph, but a false statement definitely does not. That is very important!" Well, I then made a statement such that she had to give me, not her autograph, but a kiss! Can you guess what statement would work? I'll tell you a bit later, but first I will tell you another way I like to win kisses from ladies. I say to the lady, "Do you believe it is possible to kiss a lady without touching her?" The lady usually agrees that it is impossible. I then tell her, "I'll bet you I can kiss you without touching you! Will you take the bet? The bet doesn't have to be about money. I mean only a friendly bet." The lady usually takes the bet. I then say, "Alright, close your eyes!" She closes her eyes, I give her a kiss and say, "I lose!"

Now for the answer to the question of what statement I could make to the lady musician such that she would have to give me a kiss. Well, the statement I made was, "You will give me neither your autograph nor a kiss!" Let us analyze this carefully. Suppose the statement were true. Then, as agreed, she must give me her autograph. But giving me her autograph would falsify the statement that she gave me *neither* her autograph nor a kiss! Hence if the statement were true, it would also have to be false, which is impossible! Thus the statement can't be true and must be false. Since it is false that she will give me *neither* then she must give me *either* (false neither is the same as true either). And so she must give me either her autograph or a kiss. But she cannot give me her autograph for a false statement, hence she owed me a kiss!

This was a pretty sneaky way of winning a kiss, wasn't it? Well, what happened next was even more interesting. Instead of collecting the kiss, I suggested we play for double or nothing. She, being a good sport, agreed. Then with another trick, she owed me two kisses, then four, then eight, and things kept doubling and escalating and doubling and escalating, and before I knew it, I was married! And I was happily married to Blanche, the lovely musician, for forty-eight years.

The statement I made which won that first kiss had to be false. There is another statement I could have made which would have to be true and such that she would again have to give me a kiss. Can the reader guess what statement that could be? There is still another statement I could have made — a particularly interesting one — such that there is no way of telling whether it is true of false, but in either case, she would have to give me a kiss! What statement could that be?

Well, here are the answers. A statement which must be true, and which guarantees me a kiss is "Either you will not give me your autograph, or you will give me a kiss." With this statement I am asserting that at least one of the following two alternatives holds:

(1) You will not give me your autograph.
(2) You will give me a kiss.

If the statement were false, then neither alternative (1) nor (2) would hold, and since (1) wouldn't hold, that would mean that she *would* give me her autograph, contrary to our agreement that she can't give me her autograph for a false statement. Thus my statement can't be false and must be true. Since it is true, this means that either (1) or (2) holds, but (1) can't hold, since she must give me her autograph for a true statement. Therefore (2) must hold, and so she must give me a kiss (as well as her autograph).

Now for the more interesting statement I could have made, such that from the statement alone there is no way of telling whether it is true of false, yet in either case, she must give me a kiss. Well, such a statement is, "You will either give me both your autograph and a kiss, or neither one." Suppose the statement is true. Then she can't give me neither, for then she would fail to give me her autograph for a true statement. Hence she must give me both. And so in this case she must give me a kiss (along with her autograph). Now consider the case that the statement is false. This means that she gives me one but not the other, either a kiss and no autograph, or an autograph and no kiss. It can't be that she gives me her autograph (since we are considering the case of the statement being false), hence she must give me the other, a kiss. In summary, it could be that the statement is true, in which case she must give me a kiss as well as her autograph, or the statement could be false, in which case she must give me a kiss but not her autograph. From the statement alone it cannot be determined whether it is true or false. It can be determined only after the lady acts. She must give me a kiss, but she has the option of giving me her autograph, which would make the statement true. If she doesn't, that would make the statement false.

Blanche was indeed a musician. She was a pianist and teacher and the head of a lively music school in Manhattan named "The Music House". Many instruments were taught there, as well as theory and composition. She had a distinguished faculty, which included the members of the Juilliard Quartet, the flutist Samuel Baron of the New York Woodwind Quintet, and the distinguished cellist Leonard Rose.

I must tell you a famous incident about Leonard Rose. He was one of a distinguished trio, with the pianist Eugene Istomin and the violinist Isaac Stern. The Istomin, Stern, Rose trio once performed at the White House. At the conclusion of the performance, the President said, "I wish to congratulate Mr. Stern and his two accompanists."

As I have already mentioned, I was married to Blanche for 48 years, until she passed away in 2005 at the ripe old age of 100! I was 14 years younger than she was.

Once at breakfast, I was in a particularly mischievous mood, and felt like teasing Blanche, and the following dialogue took place:

Raymond: Is *no* the correct answer to this question?
Blanche: To what question?
Raymond: To the question I just asked. Is *no* the correct answer to that question?
Blanche: No, of course not!
Raymond: Aha! You answered "no", didn't you?
Blanche: Yes.
Raymond: And did you answer correctly?
Blanche: Why, yes!
Raymond: So no *is* the correct answer to the question!
Blanche: Yes.

Raymond:	So when I asked you whether it was, you should have answered *yes*, not *no*!
Blanche:	(After more thought.) I guess you're right. I should have answered *yes*.
Raymond:	No, you shouldn't have!
Blanche:	What?
Raymond:	If you answer *yes*, you are affirming that *no* is the correct answer, in which case you should give the correct answer *no*, not *yes*!
Blanche:	You're confusing me!

Fortunately Blanche didn't divorce me for this! It's pretty difficult having a mischievous logician for a husband, isn't it? The following dialogue I once wrote shows why wives don't want their husbands to be overly rational:

Wife:	Do you love me?
Husband:	Well, of course! What a ridiculous question!
Wife:	You don't love me!
Husband:	Now what kind of nonsense is this?
Wife:	Because if you really loved me, you couldn't have done what you did!
Husband:	I have already explained to you that the reason I did that was *not* that I don't love you, but because of such and such.
Wife:	But that such and such is only a rationalization. You really did it because of so and so, and this so and so would never be if you really loved me.

Etc., etc.!

Next Day

Wife:	Darling, do you love me?
Husband:	I'm not so sure!
Wife:	What!
Husband:	I thought I did, but the argument you gave me yesterday proving that I don't was pretty good.

110

It is also good if a husband is not overly pedantic, as the following poem by Ogden Nash, entitled "The Purist", will reveal:

I give you now Professor Twist,
A conscientious scientist.
Trustees exclaimed, "He never bungles!"
And sent him off to distant jungles.
Camped on a tropic riverside,
One day he missed his loving bride.
She had, the guide informed him later,
Been eaten by an alligator.
Professor Twist could not but smile.
"You mean," he said, "a crocodile."

Blanche was born in Belgium. Her musical ability was recognized at a very early age. I must now tell you a very beautiful and moving story. During World War I, when Blanche was a child, the Germans occupied Belgium and homes were forced to house German soldiers. A German officer was housed with Blanche's family. Fortunately he was quite musical and loved to hear Blanche's playing. For hours he sat beside her listening to her practicing. Near the end of the war, he had a furlough and went back to Germany. When he returned, he brought Blanche beautiful gifts of rare and expensive music editions! [When I told this to a friend, he said, "In music, there are no enemies!"]

Another person who liked to tease Blanche was Melvin Fitting. I have already told you some things about Melvin, and now I want to tell you some more. He was a frequent visitor to our house in the upper Catskill Mountains. Indeed, he was married to his first wife Greer in our house. They were married by a Justice of the Peace. When the Justice of the Peace arrived at our home, my female dog was on the porch mating with a male dog. The Justice was quite embarrassed, and Blanche, seeing this, said, "Perhaps you had better marry them first!" He was so rattled that when he was supposed to say, "Do you, Greer, take this man Melvin to be your lawful wedded husband?" he said, "Do you, Greer, take this man Felix to be your lawful wedded husband?"

Mel and Greer slept over at our house that night and for several other nights. During the whole interval, our female dog was in heat and male dogs were constantly howling through the nights. It was during that time that Greer became pregnant with her daughter Miriam.

Miriam was really a chip off the old block! When she was six years old, she and her father Melvin were having dinner at our house, and Melvin said, "That's no way to eat, Miriam!" Miriam replied, "I'm not eating Miriam!" Pretty clever for a six-year-old, no? Mel and Greer had a second child, Rebecca. Some years later, Mel and Greer got divorced.

Mel's second wife is named Roma.

Mel has a great sense of humor. Once at our house, someone complained that it was cold. Mel said, "As it says in the Bible, many are cold, but few are frozen."

Once when Mel and I were driving through town, Mel said, "What are all these signs I see advertising slow children?" On the same trip, we passed a sign advertising A-frame houses. Mel said, "A friend of mine recently built a D-frame." I asked what a D-frame was, and Mel said, "I don't know, but he sued the architect for deframe nature of character."

Once when Mel slept over at our house, the next morning at breakfast he showed me the following poem he had just written:

> I want to write a poem
> On a piece of paper.
> On its surface, that is,
> Not on its essence.
> Now, to extract the essence of paper,
> May be profound,
> But it makes it mushy
> And hard to write on.

Several years earlier, when Mel was writing his doctoral thesis with me, he had to be away for a summer. I wrote him a letter which concluded with, "And if you have any questions, don't hesitate to call me collect and reverse the charges." [Get it?]

In the year 2011 or 2012, Mel sent me a copy of an excellent book on mathematical logic he had written. I then sent him a message, which I also emailed to many of our mutual friends:

My former student, now Professor Melvin Fitting, recently sent me an excellent book he had just written, and I wrote him the following, which I believe will amuse you. [We had just seen an excellent performance of *Hamlet* together.]

Dear Mel,

Sending me your book was both extremely kind and extremely foolish. Why kind? The answer is obvious. Why foolish? Well, did it not occur to you that I might steal some of your excellent ideas for my next book? Will I really steal them? It's difficult for me to decide.

TO STEAL OR NOT TO STEAL, THAT IS THE QUESTION:
WHETHER 'TIS NOBLER IN THE MIND TO REFRAIN,
OR YIELD, AND THUS INCUR
THE SLINGS AND ARROWS OF REMORSE.
TO DIE, TO SLEEP;
TO SLEEP: PERCHANCE TO DREAM,
TO DREAM OF RETRIBUTION YET TO COME!
AYE, THERE'S THE RUB!
OH HORRORS, CAN I NOT REFRAIN?
ANGELS ABOVE, HELP MY ANGUISHED SOUL!
OH, CURSED ILL
THAT I WAS EVER BORN TO STEAL!

Well, Melvin, I finally decided that I WILL steal all your excellent stuff.

OH, MY DEED IS DONE AND IS MOST FOUL!
ITS SMELL GOES UP TO HEAVEN.
THAT TEMPTATION SHOULD HAVE GOT THE BETTER OF REASON,
'TIS A HORROR NOT TO BE ENDURED.
AND YET, I CANNOT GIVE UP —
THE FRUITS OF MY VILE DEED.
TO STEAL FROM STUDENT AND LOYAL FRIEND,

WHAT WORSE EVIL COULD THERE BE?
OH HEAVEN, IS THERE NO HELP
FOR THIS TORMENTED SOUL?
AH, BUT LIKE CLAUDIUS,
MY WORDS GO UP, My thoughts go down.

Now that I have decided to steal, the next question is whether I should give you credit. This is a more difficult matter. Of course I am open to bribery. I somehow think that a kiss from your lovely wife Roma may be relevant to all this.

I sent a copy of the above to another former student, Bruce Horowitz, who replied:

Dear Raymond,

Cute! Now, over the years, I've noticed that most debts owed to you can be paid by a kiss from a fair maiden. Given your paraphrasing of Hamlet, I will paraphrase Richard III:

A kiss, a kiss! My kingdom for a kiss!

Also, Melvin replied to my message:

Dear Raymond,

It is said, somewhere, "Neither a borrower nor a lender be." Since it doesn't mention stealing, I guess that's OK.

Best, Mel

Now I must tell you of a very remarkable incident. In Windham, N.Y., which is about a 20-minute drive from my house in Elka Park, is a very lovely restaurant named *The Chalet Fondue*. It changed hands about a year ago, but the former owners, Uta and Kurt, were from Bremen, Germany. Blanche and I ate there frequently, and the owners were very fond of Blanche. Well, in December of 2005, about a month before Blanche passed away, her son and several of Blanche's grandchildren were at our house. When dinner time came, I phoned in to the Chalet Fondue an order of food for several hundred dollars, and said that I would drive over and pick it up. Well, I started out driving, but it was snowing so badly that I soon turned back and phoned the restaurant to explain that it was snowing too badly for me to come, and so I had to cancel the order. Uta told me, "We have a four-wheel drive, and snow is no problem, and so we will bring the food over to you." A short time later, Uta and Kurt arrived with the food *and wouldn't let me pay for it*! Isn't that amazing?

* * *

I must now tell you of a most remarkable recent sequence of events. Once I was rummaging through some old correspondence and came across the following letter, written to me four years ago, which might be of historical interest, since it was written by a descendent of a well-known public figure, whose identity you will soon know, and which I knew only towards the end of the letter.

Dear Dr. Smullyan,

Usually I would not write a letter like this. Some would call this a fan letter, but I think that's not exactly right. I am a reader of yours. I love finding a book of yours on a bookshelf. That seldom happens anymore since I live in a small town in Utah. I usually have a local bookstore search suppliers periodically to see what new books you have written.

For years one or another of your books accompanied me on backpacking trips in the mountains and canyons of Utah. In the evenings I would read them to my son or daughter. We would talk about the ideas and I could see the light of logic growing behind their eyes. Sometimes we tried to solve a puzzle as we hiked.

I have been reading *Who Knows*. Towards the end you say, "… that I believe in some kind of afterlife — what kind I do not know!" I have a question about this. I do not mean to be argumentative. Evolution explains how we came to be. I think the power of evolution to explain human life, thoughts and emotions grows all the time. My question is, what advantage would an afterlife bestow on a living being? Which of all the creatures on this earth would have evolved to their current form because or necessarily with an afterlife?

Perhaps an afterlife is bestowed upon us by some mysterious random power flowing through the universe or by some benign and loving force with a responsibility to sentient beings who can speculate on death.

I have sometimes thought the power of belief and faith could prop up a man before other men. Such a man would have an advantage for living and reproduction in human society. In that case, it is the fervency of faith (or perhaps the religious impulse) rather than an actual afterlife that would give a man an advantage in passing on his genes.

As the great grandson of Brigham Young I can say such an advantage gets diluted very quickly.

Whatever awaits us after death, I hope it is as pleasurable as the pleasure your books have given me.

Sincerely,
Wayne Hoskisson

Isn't that remarkable? The writer really is the great, great grandson of the famous 19th century Mormon leader Brigham Young! We subsequently had a good deal of very interesting correspondence, and I have saved all the letters. I sent him some of my videos, including the documentary I made of the life of my wife Blanche, about which he wrote: "I was impressed that you were especially drawn to a couple of photographs when Blanche gave a captivating glance at the camera. I can remember some moments like those in my life. They are the seeds of joy."

Then he wrote, "One of the qualities I found in your books was a sense of joy. You write about logic, but in a way that allows a human quality to poke through, as though a judge would bring a vase of tiger lilies to court to grace his bench."

Now, we had decided that if he ever came east, he would visit me. Well, after reading through this correspondence, I felt a pang of remorse at not having contacted him these last four years, and so I went to the phone to call him. To my horror, I was told that that phone number is no longer in service! My heart was in my mouth! I was afraid that maybe he had passed away, and so rushed to my computer and sent him an email. I got a message that the email could not be delivered!! Of course this reinforced my fears. Then I went to dinner with some very good friends, and after reading them several of the letters (which amazed them and which they very much enjoyed), I told them of my unsuccessful attempts at contacting him and that I wondered whether there was any way I could find out whether he was alright. Then one of my friends suggested the obvious thing, which I was stupid enough not to have thought of — namely Google! I went home and did this, and sure enough his name came up in the same town, but with a different address and email. I then emailed him the following: "If you are the Wayne Hoskisson that I know, please contact me." Five minutes later the phone rang, and there he was on the line, alive and kicking! My God, what a relief!

9 | The Piano Society

A few years after my wife passed away, I received a letter from a lady violinist in Canada named Claudia Shaer, who wrote that she was a fan of my puzzle books and was interested in mathematics and music, and was particularly delighted to learn that I was a concert pianist. She wanted some general advice from me. I phoned her and we arranged that she come and visit me, which she did, and we became very close friends. Claudia is a graduate of the Juilliard School and of SUNY Stony Brook. She has taught and performed throughout the United States, Canada, Europe and China, where she was named guest professor at the Guangxi Arts College of Nanning, China. Claudia is also an advocate of contemporary music. She has performed at the Lucerne Music Festival under the baton of Pierre Boulez.

When Claudia visited me, she played the famous Bach Chaconne for me and some friends. I enjoyed it more than any other performance I ever heard! She and I also played some four-hand piano music. She was a much better sight reader than I am!

It was Claudia who introduced me to the Piano Society,[4] an event that virtually changed my life! This Society is a wonderful international organization on the Internet, in which both amateur and professional pianists can upload their recordings and anyone can download them at no cost. The level of playing is extraordinarily high. Some of the amateurs are at

---

[4] http://pianosociety.com

least as good as many of our best professionals. One visitor once remarked that on at least three occasions, the performance of a piece he heard on the site was better than any other they had heard in their lives. And I can personally testify that the performances of the Chopin Fantasie Polonaise and the B Minor Sonata are the best I have ever heard.

The Society offers several other features. Composers can upload their recordings, and visitors can view and hear some lovely videos by participants as well as by famous pianists who are not members of the Society. In addition to some of my CDs that I uploaded there, I have several videos which combine my piano playing with my photography. In one of them, consisting of a Schubert sonata, in one of the movements, instead of using my photography, I use paintings of a Piano Society member, Julia Froschhammer, who is both an excellent pianist and a superb painter.

Julia's paintings can be seen online at http://juliafroschhammer.com/juart. html. The picture you can see of her there will help you understand why I wish I were sixty years younger! Julia has had an interesting career. She was born in Germany and now lives in Switzerland. She began piano lessons at the age of six with her father, the pianist and composer Felix Froschhammer. Her whole family is very musical. Her brother Felix is an excellent violinist, and the two often concertize together. They won the only prize in the category of violin and piano at the International Wolfgang Jacobi Competition for Chamber Music of the 20th century.

Coming back to Claudia Schaer, the way she met me and introduced me to the Piano Society had best be described in her own words (as given in a book I will soon describe):

It was I who introduced the instigator of this book to the Piano Society, and the story is one of remarkable coincidences. As a violinist, I knew of Raymond Smullyan's books thanks to the teacher of a logic class I took at Columbia University to satisfy my liberal arts requirement at Juilliard. When beginning my doctorate at Stony Brook University, I decided to let his books keep me company on the train between Manhattan and Stony Brook, as they are written in such a gentle and sensible vein, and I enjoy the puzzles very much.

When I then found out that Raymond had also been a concert pianist, I decided to write to him in the hopes that he might be able to offer some suggestions for how to balance musical, mathematical, and philosophical interests, without being thoroughly overwhelmed.

He in turn invited me to meet him, and, as we continued correspondence, I couldn't resist forwarding to him the link I'd found to the piano roll of Hoffman playing Beethoven's "Rage over a Lost Penny" on the Piano Society website, which I happened to come across when I was surfing the net one evening looking for a recording of Liszt's *Bagatelle Without Tonality* (for a paper for my "New Approaches to Analyzing 19th-century Music" class).

As they say, the rest is history!

Claudia Schaer

One lovely aspect of the Society is the Forum, which includes the "Audition Room", a delightfully lively place where visitors can hear the most recent uploads and make comments about the performances. The members make the most helpful suggestions, and all in a very friendly spirit. There is a wonderful feeling of comradery. Several members have stated that their entrance to the Society has been like finding a new family. That was certainly the case with me! I have had all sorts of wonderful email conversations with many of the members. Let me now tell you a particularly delightful incident.

One of our members is the excellent pianist Hannah Shields. [If you ever see a photograph of her, you will find another reason I wish I were 60 years younger!] She has performed throughout the United States both as a soloist and as a chamber player. She has appeared at Carnegie Hall as a soloist under the baton of Itzhak Perlman and has performed at Weil Hall, Alice Tully Hall, the Seattle Opera House and Boston's Jordan Hall. She has been a top prize winner in many solo competitions, including the 2002 Kosciuszko Chopin Competition.

She once posted a Beethoven sonata in the audition room. After praising it, I had the audacity, gall, nerve, effrontery, to write "Love" before signing my name! Well, I was then afraid I had overstepped my bounds, but to my great relief, she soon sent me a private message, "Dear Mr. Smullyan, I was so pleased to read that you enjoyed my playing — especially because I have had the pleasure of reading several of your books!"

A lovely email interchange ensued in which I told her that I would have become a professional pianist, had I not developed tendonitis in my right arm. She told me that she had a Chinese violinist friend Melody Chan, who had a similar problem with her left hand, and had to stop playing for a while. She told me that Melody is now a graduate student at Princeton working for a Ph.D. in mathematics. Princeton! Where I got my degree in mathematics!

The upshot is that Hannah and Melody visited me one weekend, and we all had a wonderful time! Melody Chan is quite remarkable — besides playing the violin and working on a Ph.D. in mathematics, she plays the piano extremely well. She is also fantastically intelligent. And it was she who had introduced Hannah to my puzzle books!

It is interesting how mathematicians can manifest their mathematical intelligence even when discussing topics that are not mathematical. Melody Chan did this in the following manner: I was reading to her and Hannah part of my book *To Mock a Mockingbird*, which is a popularization of a field known as *combinatory logic*, and I chose to use *birds* as the mathematical objects under consideration.

Inspector Craig (the hero of many of my books) was looking for the Master Bird Forest, and after many adventures, finally arrived. When Craig reached the entrance, he saw an enormous sign hanging over the gate:

THE MASTER FOREST
ONLY THE ELITE ARE ALLOWED TO ENTER!

"Oh, heavens!" thought Craig. "I have no idea if they will let me in. I've never thought of myself as elite; in fact, I'm not quite sure I know what the word really means!"

At this point, an enormous sentinel blocked his way.

"Only the elite are allowed to enter!" he said in a terrible voice. "Are you one of the elite?"

"That depends on the definition of 'elite,'" replied Craig. "How do you define an elite?"

"It's not how *I* define it that counts. It's how Griffin defines it."

"And who is Griffin?" asked Craig.

"Professor Charles Griffin — he is the resident bird sociologist of this forest, and he's boss around here. It's *his* definition that counts!"

"Then what is his definition?"

"Well," replied the sentinel in a softer tone, "his definition is a very liberal one. He defines an elite as anyone who wishes to enter. Do *you* wish to enter?"

"Of course!" said Craig.

"Then by definition, you're an elite and are free to enter."

"That's a relief," thought Craig, as he wended his way to meet Professor Griffin. "I wonder why he instituted such a strange rule, which in fact doesn't exclude anybody!"

Well, after Craig had had many adventures in the Master Forest, came the sad day he had to leave. "This vacation has been like an idyllic dream," he thought, as he reached the exit gate — also the entrance gate. "I really must visit this forest again!"

"Only the elite are allowed to leave this forest!" said an enormous sentinel who blocked his way. "However, since you have entered the forest, and only the elite are allowed to enter, then you must be one of the elite. Therefore you are free to leave, and God speed you well!"

"This is one ritual I will never understand," thought Craig, as he shook his head.

Well, when I had read this to Melody and Hannah, Melody looked thoughtful, and finally said, "That means that if a person is born in the forest, he might never be able to leave!"

To me, that was a brilliant thought. Unfortunately, I never saw Melody again after that lovely weekend. But I did see Hannah again, in a musical gathering in Boston which I organized for several pianists. The event was videoed, and can be seen on YouTube.

Sometime, I believe it was in 2008, I realized that the Piano Society needed funds, and so I had the bright idea (if you will excuse my boasting) of making a book of profiles of the pianists of the Society in which they tell the story of their lives, and give their views on music and art and other things. All moneys earned by sale of the book would go to the Piano Society. The book came out beautifully! This is largely due to the expert formatting and editing of the physicist and pianist Dr. Peter Bispham of England (also a member of the Society), who also designed the beautiful front cover, using photographs of paintings of Julia Froschhammer. [This is the book referred to in Claudia's account of how she met me and introduced me to the Piano Society. The book is available at Lulu Press and also at Amazon, and possibly also from the Piano Society.]

One of the things that makes the book so interesting is the remarkable diversity of the professions of the pianists! The book is entitled *In Their Own Words — Pianists of the Piano Society*. Of all the books I have published, I am especially proud of this one, even though I am not the author, but only the instigator, compiler and editor, and one of the many contributors (yes, my profile is also included).

The book got the following reviews:

"The volume, In Their Own Words, presents a fascinating compendium of the world of the piano. The Piano Society is an important member of the musical community."

David Dubal
Internationally-known pianist, professor at the Juilliard School, painter, broadcaster and well-known author of many books on music and musicians.

124

"This is an interesting book, in which a number of pianists, professionals or amateurs, tell of their work, their tastes, their life. It brings a fresh glance at a profession considered as superhuman."

Thérèse Dussaut
Pianist and Emeritus Professor of the Piano Department
at the Conservatoire Toulouse and jury member of many
international competitions, including the Tchaikovsky in Moscow.

"I think your book is very interesting and an important contribution."

Al Goldstein
Founder of Pandora Records.

"From this important book, it is most rewarding for me to learn about the efforts and accomplishments of the Piano Society, whose members, in a world where most entertainment and pastime is spent in front of a screen or plugged into earphones, devote a great deal of their time playing and practicing the piano with its immense literature. Such spiritual, mental, and physical pleasure cannot be obtained in any other way."

Lionel Party
Harpsichordist and professor at the Juilliard School and
recording artist for the Musical Heritage Society,
Smithsonian, Orion Records, and many other labels.

"This is a fascinating and skillfully-produced book of profiles of pianists — both amateurs and professionals — of the Piano Society, who charmingly tell about their lives and views on music and other arts. A must for those interested in the daily lives of practicing musicians."

Lisa Kovalik
Pianist and professor at the Juilliard School.

The Forum has many departments, one of which is the Audition Room, which I have already discussed. Another is called "General", which is open only to members, but membership is free. Soon after I joined the Piano Society, I posted the following musical definitions:

Musical Definitions

Group 1 [Some of these are original to me, others are already known, and others are variations of mine on known ones.]

Air. That which must be inhaled by singers before starting to sing.

Airs. Peter Pears
Needn't give himself airs.
He has them written
By Benjamin Britten.

Analysis. A musical dissection which, like an autopsy, results in a lifeless dismembered corpse. *Musical Analysts* are so-called, because of their need to be psychoanalyzed.

Applause. Noise made by audiences, usually between movements, to signify that they have heard enough!

Bach. He was the world's greatest composer, and so was Handel.

Bang Box. A less polite, but more accurate, name for a piano.

Barcarole. A sound made by a rolling dog. A famous Barcarole was composed by one who himself would often-bark.

Bass. A musical fish with a low voice.

Bassoon. A wind instrument played by buffoons and suitable for baboons.

Beethoven. A composer who had hearing problems, hence wrote music that was very loud.

Bronze. The last name of a composer who wrote four great symphonies. His first name was Johannes.

Bram. The composer of Brahms' symphonies.

Carmen. Parking attendants.

Cellist. One who greets one with "Cello!"

Choir. A musical organization invented by Handel to perpetuate his Messiah.

Composers. Those who write music during their lifetimes and later decompose.

Clarinet. A wind instrument named in honor of a Russian girl Clara who played so much that finally someone said to her, "Clara! Nyet!!"

Conductor. One whose function is to foul up an orchestra.

Crook. A concert manager.

Discord. Not to be confused with datcord.

Encore. An extra piece performed "off the cuff" that has been well-prepared for several weeks.

English horn. An instrument so named because it is neither English nor a horn. Not to be confused with the French horn, which is German.

Exposition. A position no longer in service.

Fermata. A fitting appendage to John Cage's 4.3 moments of silence.

Flute. A flying lute, which was discovered by a fluke.

Glissando. A musical equivalent of slipping on a banana peel.

Harpsichord. A precursor of the piano, and cursed ever since!

Klavierstuck. A German term for a piano stuck in a narrow doorway.

Leirmann. A German term usually translated as "organ grinder". I think this is a mistranslation. To me, a leirmann is a man who doesn't tell the truth.

Lute. An instrument used to accompany songs that are lewd.

Madrigal. A Rigal that has gone mad.

Mezzo-Soprano. A half-hearted soprano.

Modes. Scales that are outmoded.

Musical Snob. One who pretends to know more about music than we pretend to.

Oboe. An instrument taught at Oberlin College, named in honor of its inventor, who was a hobo.

Offenbach (frequently pronounced *Offenbark*). A dog that barks frequently.

Opera. A performance that is viewed with opera glasses.

Piano. An instrument operated by depressing the keys and the spirits of the listeners.

Piano Tuner. A species of musical fish.

Pianist. One who knows how to get rid of unwelcome guests.

Podium. A raised platform for conductors to make them feel more important than they really are.

Posthumous. Works of art produced by the artist after he is dead.

Refrain. What most performers had best do.

Repertoire. The musical knowledge of General Ulysses S. Grant, who said, "I know two pieces. One of them is Yankee Doodle and the other isn't."

Serial Music. Music that is different from breakfast music, which is cereal music.

Sexaphone. An instrument known for its pornography.

Spinet. As one woman who had just read a biography of Bach said to a friend, "Did you know that Bach had twenty wives and at night would go up in the attic and practice on an old spinster?"

Squeak Machine. The highest pitched of the stringed instruments, which is usually accompanied by a Bang Box. Beethoven wrote ten squeak machine and bang box sonatas.

Stop. What organists should do (cf. *refrain*).

Triangle. An instrument played by squares.

Trill. A musical form of an epileptic fit. They say that since the advent of modern music, much of the trill has gone out of life.

Trumpet. A musical instrument that is supposed to trump all others, but in reality is fit only for strumpets.

Variations. The kind of music written by composers who cannot make up their minds.

Viol. An instrument so named because of the vile sounds it produces. Hearing it often makes one violent. One who plays it wrongly is called a violator.

Violin (pronounced *viol-in*). A viol that is not out.

Virginal. A keyboard instrument similar to the harpsichord, so-called because of the sorts of ladies who were supposed to play it in the 16th century — and if you believe that, you'll believe anything.

Group 2 [None of these are mine. They are excellent!]

Adagio fromaggio. To play in a slow and cheesy manner.

A la regretto. Tempo assigned to a performance by the conductor after it is panned by the local music critics.

Al dente con tableau. In opera, chew the scenery.

Allegro con brillo. The fastest way to wash pots and pans.

AnDante. A musical composition that is infernally slow.

Angus Dei. A divine, beefy tone.

Antiphonal. Referring to the prohibition of cell phones in the concert hall.

A patella. Unaccompanied knee-slapping.

Appologgiatura. An ornament you regret after playing it.

Approximatura. A series of notes played by a performer and not intended by the composer, especially when disguised with an air of "I meant to do that."

Approximento. A musical entrance that is somewhat close to the correct pitch.

Bar line. What musicians form after a concert.

Basso continuo. The act of game fishing after the legal season has ended.

Basso profundo. An opera about deep sea fishing.

Cacophany. Composition incorporating many people with chest colds.

Concerto grosso. A really BAD performance.

Coral Symphony. (See: Beethoven — Caribbean period.)

D. C. al capone. You betta go back to the beginning, capiche?

Dill piccolo. A wind instrument that plays only sour notes.

Diminuendo. The process of quieting a rumor in the orchestra pit.

Eardrum. A teeny, tiny tympani.

Fermantra. A note that is held over and over and over and … .

Fiddler crabs. Grumpy string players.

Flute flies. Gnat-like bugs that bother musicians playing outdoors.

Fog horn. A brass instrument that plays when the conductor's intentions are not clear.

Frugalhorn. A sensible, inexpensive brass instrument.

Gaul blatter. A French horn player.

Grace note. The I.O.U. you deposit in the church collection plate when you're out of cash.

Ground hog. Someone who takes control of the repeated bass line and won't let others play it.

Kvetchendo. Gradually getting ANNOYINGLY louder.

Opera buffa. A musical stage production at a nudist's camp.

Pastorale. Beverage to drink in the country when listening to Beethoven with a member of the clergy.

Pipe smoker. An extremely virtuosic organist.

Pizzacato. The act of removing anchovies from an Italian dish with short, quick motions and tossing them to a nearby awaiting feline friend.

Placebo Domingo. A faux tenor.

Rights of Strings. Manifesto of the Society for the Prevention of Cruelty to Bowed Instruments.

Rubato. Cross between rhubarb and a tomato.

Schmaltzando. A sudden burst of music from the Guy Lombardo band.

Spritzicato. Plucking of a stringed instrument to produce a bright, bubbly sound, usually accompanied by sparkling water with lemon (wine optional).

Tempo tantrum. What a young orchestra is having when it's not keeping time with the conductor.

Toiletto. The effect on the human voice of reverberations in small rooms with ceramic tiles.

Trouble clef. Any clef one can't read, e.g. the alto clef for pianists.

The next post was a music quiz.

Music Quiz, by Raymond Smullyan

[All of these items are original, except for the three starred ones, the first (and best) of which is due to Sylvie Degiez.]

Feline Composers

(1) Name some composers with feline tendencies.
(2) What composer is doubly feline?
(3) What composer is feline and nothing more?

Answers

(1) Kat-chaturian
Milhaud (pronounced "Meeoh")
De Pussy
Modeste Meowski
Gustave Meowler
Giacomo Pouse-ccini
Harry Purr-cel
Claws Monteverdi
Benjamin Kitten
(2) Claws de Pussy
(3) Ferruccio Pouse-Only

Canine Composers

(1) Name some composers with canine tendencies.
(2) What two composers are not always canine, but frequently so?
(3) Name an operatic composer with canine tendencies.
(4) Name a violinist composer with canine tendencies.
(5) What canine-like composer has a beautiful voice?
(6) What canine-like composers resemble a happy dog wagging its tail?
(7) What canine-like composer is doubly canine and also artistic?
(8) What pianist is more canine when he is at home?

Answers

(1) Johann Sebastian Bark
Ludwig van Bark-hoven
Hugo Wolf
Franz Joseph Hyena
Bela Bark-dog
Woof-gang Amadeus Mozart
(2) Jacques Often-bark and Ludwig van Bark-often
(3) Richard Bark-ner
(4) Nicolo Bark-anini

(5) Arnold Shoen-bark

(6) Nicolo Wag-anini and Richard Wag-ner

(7) Wolf-gang Amadeus Mutts-Art

(8) Wilhelm Bark-house

Watery Composers

Name some watery composers, or other musicians reminding one of liquids.

Answers

Pala-Stream-er

Sir Thomas Beach-um

Edward Mac-Towel

Brook-ner

Al-Bay-nitz

Max Brook

Frederik Ku-Lake

Ludwig van Bay-thoven

Sea-Bay-lius

Meyer-beer

Wine-iavski

Frederic Cho-Pond

Franz Joseph Hydrant

Bovine Composers

Name some bovine composers.

Answers

Tchai-Cow-ski

Rimski-Cow-sakov

Hans von Bull-ow

Ferraccio Bull-soni

Vincenzo Bull-ini

Artur Schna-Bull (he also composed)

Composers with Colds

(1) What composers have colds?
(2) What composer most certainly has a cold?

Answers

(1) Tchai-Cough-ski
Sergei Pro-Cough-iev
Rimski Korsa-Cough
(2) Rimsi-of-course-a-Cough

Composers and Money

(1) What composer gambles a lot?
(2) What composer wins a lot?
(3) What composer makes frequent bets?
(4) What composer has money?
(5) What composer has enough money?

Answers

(1) Domenico Scarlottery
(2) Win-iavski
(3) Ludwig van Bet-often*
(4) Rach-money-nov
(5) Rach-money-enough

Miscellany

(1) What composer belongs in a delicatessen?
(2) What composer had hardly enough cereal?
(3) Name some musicians related to time.
(4) What composer is prepared to write on a blackboard?
(5) What composer exhibits kitchen ware?
(6) What composers and other musicians are over-heated?
(7) What composer never goes barefoot?*
(8) What composer always goes barefoot?*
(9) What composer is angelic?
(10) What conductor has an elephant-like quality?

(11) What pianist is jewel-like?

(12) What two composers should you take with you when you buy groceries?

(13) What composer was physically abusive?

(14) What pianist frequently went into the water?

(15) Name a hybrid composer.

(16) Name a sleepy hybrid composer who has Russian and Irish tendencies.

(17) What composer reaches green heights?

(18) What composers sound when struck?

(19) What composer has gangster-like tendencies?

Answers

(1) Al-Baloney

(2) Hector Barely-oats

(3) Leopold Hour
Vladimir Hour-witz
The SECOND son of Bach
Ponchielli, who wrote "Dance of the Hours"
Chopin, who wrote a Minute waltz

(4) Got-Chalk

(5) Frederic Show-Pan

(6) De Falla (pronounced de fire)
Ludwig van Burn-oven
Leonard Burn-stein
Van Cli-Burn

(7) Robert Shoe-mann

(8) Franz Shoe-bare

(9) Cherubini

(10) Arturo Tusk-anini

(11) Artur Ruby-stein

(12) Chopin, Liszt (Shopping List)

(13) Gustav Maul-er

(14) Edwin Fish-er

(15) Rimski Tchai-Korsakoff-ski

(16) Modeste Mc-Snore-ski

(17) Claude Mountain-verdi
(18) Bell-ini
      Hans von Bell-ow
      Feraccio Bell-soni
      Pachel-Bell
(19) Buxte-hoodlum

Other Musicians

(1) What pianist is in a hurry?
(2) What pianist has feline characteristics?
(3) What pianist keeps returning home?
(4) What violinist brings tears to one's eyes?
(5) What violinist has strong seizures?
(6) What violinist is helpful in seating people in a theater?
(7) What violinist is useful for sewing?
(8) What cellist likes to play with a certain toy?
(9) What pianist will help you get home?

Answers

(1) Myra Haste
(2) Dinu Lu-kitty
(3) Wilhelm Back-House
(4) Fritz Cry-sler
(5) Jascha High-Fits
(6) Usher Heifetz
(7) Thimble-ist
(8) Yo yo Ma
(9) Will-help Back-House

Non-Musicians

(1) What painters are always on the move?
(2) What painter chooses his donkey?
(3) What painter remembers his aunt?
(4) What author resembles a weapon being rapidly moved back and forth?

(5) What author has difficulty in his speech?

(6) What New England author is well-known?

(7) What New England author is chesty?

(8) What author is clearly related to a banana?

Answers

(1) Van Go and Go-gauin

(2) Pablo Pick-Ass-o

(3) Remember-ant

(4) William Shake-Spear

(5) Ernest Hem-n-Haw

(6) Ralph Waldo Eminent

(7) Henry Thorax

(8) Eugene O'Peel

Some Other Questions

(1) What are caution children?

(2) What makes dogs religious?

(3) When a book is tired, why should it be sent to Rumania?

(4) In the Middle Ages, a Buddhist came to a Hungarian city and tried to convert everyone to Buddhism. He was very aggressive and made a real nuisance of himself. He was subsequently known as what?

Answers

(1) I don't know, but while driving I saw a sign on the road with those two words.

(2) When they become Dog-matic.
When they obey Canine Law.
When two of them stand at right angles to each other, they become ortho-dogs.

(3) To give the book a rest.

(4) He became known as the Buddha-pest!

Next, I started posting in an area of the Piano Society website entitled *Puzzles, Jokes, Anecdotes and Thoughts.* I now wish to give you a good sense of this interactive department. Postings from the forum that are listed in this chapter may be shortened, joined into one post (when the posts are mine), and otherwise slightly edited. Many postings have simply been skipped. For instance, if you find the solution to a puzzle immediately after the puzzle was posed, you can assume that sometimes a good number of posts actually occurred between the posting of the puzzle and the final discussion of its solution. In a post of mine as given here I will sometimes just mention the names of the members of the forums who gave a correct solution to a puzzle rather than show you all their correct solutions. Occasionally I will break out of the stream of what occurred in the forum to summarize or comment on a number of posts. When I do break out of the flow of the forum to "speak from today," I will enclose what is said in curly brackets "{}" to avoid confusing you. But I will try to retain for you much of the sense of the interaction in this forum that gives me so much pleasure.

From The Piano Society Chronicles

pianolady (Dec. 14, 2007): Our resident Puzzle Master and Humorist, Raymond, has more funny things to tell us. On a mostly daily basis, he will post a puzzle, joke or anecdote that will surely bring a smile or laugh to everyone. Here is the first one. Enjoy!

R. (*Raymond*): I love the story of Anton Rubinstein, who had a fabulous technique, but was extremely careless in his concerts, hitting enormous numbers of wrong notes. Nevertheless, his playing had so much fire and imagination, that his audiences loved it, despite the wrong notes. After one concert, Letchetitsky said to him, "You must really have a fabulous technique to be able to mess up the last movement that way!"

pianolady: Hahaha, Raymond — that one cracks me up!

R.: Speaking of wrong notes, the 20ᵗʰ century pianist Edwin Fischer (one of my favorite pianists, by the way) was known for his many wrong notes. Well, the pianist Clara Haskil and a friend were in a railway carriage discussing Edwin Fischer, and how terrible it was that he played so many wrong notes. Opposite them was Edwin Fischer, whom they did not recognize. When the train stopped, Edwin Fischer asked them, "Would you please get my suitcase from the rack. It is very heavy, because it is full of my wrong notes!"

Mark Twain defined the German language as the language in which all the verbs come in the second volume. Incidentally, Mark Twain once told a friend, "When I was in Berlin I went to the opera. I enjoyed it too, in spite of the music."

nathanscoleman: … A few months ago I had Wagner in the CD player … really loud, of course … my four-year-old came in and yelled "that's a lot of screaming, daddy!" hehe

R.: Do you know the difference between a physicist and a mathematician? Well, the following test will tell which type each is: Consider a cabin in the woods with an unlighted gas stove, an empty pot and a faucet with cold running water. What steps would you take to get a pot of hot water? Just about everyone answers that they would first pour cold water from the faucet into the pot, then light the stove and then put the pot on the stove. Well, so far, mathematicians and physicists are in agreement, but now comes the crucial test. This time the conditions are the same, except that now you have a pot already filled with cold water. How would you now get a pot of hot water? The usual reply is to light the stove and then put the pot of cold water on the stove. Well, this is the response of one who has the temperament of a physicist. A mathematician would dump out the water from the pot, reducing the problem to the previous case, which has already been solved.

A more dramatic version of the test is this: We are given a building on fire, a hose and a hydrant. How would you put out the fire? Obviously, one would seem to want to connect the hose to the hydrant, turn on the water, and then put out the fire. Now supposed the conditions are the same as before, only in this case the building is not on fire. How should you now put out the fire? The physicist would do nothing, whereas the mathematician would set the building on fire, reducing the problem to the previous case, which was already solved.

Do you know the difference between a mathematician, a physicist, and an engineer? Well, there were three such professors who had adjacent offices at a university. One day they had lunch together and came back to their offices. They all took out their pipes and started smoking. Then they each dumped their hot ashes in their wastebaskets, which contained paper and promptly caught fire. The engineer computed approximately how much water was necessary to quench the fire, took approximately that amount, and put the fire out. The physicist computed the upper and lower limits of the amount of water that was necessary, took the average, and put the fire out. The mathematician, using far more refined and sophisticated techniques, computed EXACTLY how much water would be necessary, and went back to work.

There are three kinds of mathematicians in this world — those who can count and those who cannot.

juufa72: Wow, Raymond must not like mathematicians.

R.: I AM a mathematician!

juufa72: It's a good thing if you can laugh at your own profession. Or is it?

R.: Juufa72, yes it is, or rather was, my profession.

Here is a riddle for you: A little girl said to her mother, "Let's go to the railroad station and meet Daddy, and the four of us will come home to dinner." The question is, why did she say FOUR instead of THREE? (I'll give the answer in my next entry.)

techneut: Because her little brother was coming along too, as is only common sense.

R.: No, the answer is that she was too young to count.

NEW QUESTION:

There are three errers in this messege — can you find them all?

techneut: Answer: Two are spelling errors — errers and messege. The third is the word "three", which should be "two".

pianolady: Wow — that was fast! You get the blue ribbon today.

R.: One more thought on yesterday's quiz. The answer again is: First the word ERRERS was misspelled. Secondly, the word MESSEGE was misspelled. Thirdly, there were only two errors, not three, hence the word THREE should have been TWO.

But doesn't this raise a paradox? If the word THREE was an error, then there were really three errors after all, which makes the word THREE correct! So was that word an error or wasn't it? I don't know the answer to this.

Please don't read this sentence!

techneut: I didn't.

R.: Is it possible for a person to have great grandchildren if none of his grandchildren have any children? Well, I know someone who has great grandchildren, and yet none of his grandchildren has a child? How do you explain that?

Answer (given in a later post): My friend has grandchildren and they are GREAT!

Another sneaky one. Two brothers Bob and Bill. Bob claims to have twice as many girlfriends as Bill. Bill says that they have the same number of girlfriends. Could they both be right? [Answer next time.]

Adam: They don't have any girlfriends. Twice zero is zero.

R.: Good thinking. The correct solution is usually the simplest one.

Now for a riddle: Which musical instrument is particularly intelligent?

{Answers given by others and the author: One respondent wrote, "The clarinet? As in, like … clairvoyant? That's the best I can come up with at the moment." Another wrote, "The theremin. It plays without even being touched." My answer was the clever chord.}

R.: Next puzzle: There is a 2-volume set of books sitting on a shelf. A bookworm starts from the first page of Volume 1 and bores its way to the last page of Volume 2. In each of the two volumes, the thickness of the pages is two inches, and each front cover and each back cover is 1/8 inch thick. How far did the worm travel?

Terez: Assuming that Volume 1 is on the left and Volume 2 is on the right (the usual manner) and that the books are arranged upright in a traditional bookshelf manner, it would be 1/4 inches.

R.: I was assuming that the books were in the normal position in the United States and Europe, where Vol. 1 is to the left of Vol. 2, which means that the *front* cover of Vol. 1 is right next to the *back* cover of Vol. 2. Clever you, Terez, you got it!

Here is another one, perhaps too simple: A certain drawer contains 24 blue socks and 24 red socks. A woman goes into the room where the drawer is, but the room is dark. What is the minimum number of socks she must take out of the drawer to be sure that she has two socks of the same color? [Answer next time.]

Anonymous: Instinctively I'll say 25 — the chances of the first 24 being all of the same colour are extremely remote, but to be certain …

R.: As some of you correctly said, the answer to the problem of the 24 blue and 24 red socks is 3, not 25! Since there are not 3 different colors, then given 3 socks, at least two of them must be of the same color. If I had asked what the smallest number of socks one must draw to the sure of getting at least two DIFFERENT colors, then the answer would be 25.

I am reminded of a joke. A man meets a friend on the street and says, "Do you realize that you are wearing one black shoe and one brown shoe?" "That's interesting," the friend says. "I have another pair like that at home."

Robert: Here is one: An electric train is traveling north at 200 kph and there is a headwind of 160 kph with southerly winds at 15 kph. At what speed is the smoke traveling when it leaves the locomotive?

pianolady: Hehe. An electric train? So no smoke at all unless the brakes are burning.

R.: I liked that one about the electric train. I never heard it before. I fell for it completely!

Here is another that I fell for: A certain boat had a metal ladder over the side with six rungs spaced one foot apart. At low tide the water hit the second rung from the bottom. Then the water rose two feet. What rung did it then hit?

R.: When I heard the puzzle about the boat, I said that the obvious answer is the fourth rung from the bottom, but it is too obvious to be correct, but I can't see what else it could be! I was then told that the correct answer is that the water hit the same rung (the second from the bottom) because the boat rises with the water! And so pianolady was right.

Here is another: How are you on arithmetic? Let me test you: In a certain small town, 13 percent of the inhabitants have unlisted phone numbers, and no one has more than one phone. One day a statistician came into the town and took 1300 names at random from the phone book — completely at random! Roughly how many of them would you expect to have unlisted phones?

R.: When I heard the puzzle of the unlisted phones, I said that the answer is 169 (which is 13 percent of 1300), but boy, was I wrong! The correct answer is zero, since the names were taken from the phone book! Yes, Chris, you got it.

Terez: Then there is the "Let's Make a Deal" problem. Anyone know that one? Pretty famous problem, though perhaps a bit too deceptive to call a riddle. Some people who teach probability still deny the answer to this one. Ray might be one of them. Let's hope not.

You're on that show, "Let's Make a Deal". You have three doors in front of you. Behind one door is a prize. Something like a shiny new car. It doesn't matter. You want it. Behind the other two doors are goats. And we're assuming you don't want those. Anyway, you get to pick a door, and you pick one. The host, however, knowing which door hides the prize, opens one of the doors — not the one you picked. He purposely opens a door to reveal a goat. Then he offers you the chance to change your choice.

Can you increase the odds of winning the prize by switching your choice to the remaining mystery door? Or do your odds remain the same? Can you explain the answer in terms of probability?

R.: One respondent wrote that when the door opened by the host reveals a goat, the problem is reduced to a two-door choice with a 50% chance of luck. You still don't know which of the 2 closed doors may hide the prize, so the door you already picked has as good a chance as the other. There is absolutely no point in now picking the other door, i.e. you cannot increase your chances.

However, the respondent was wrong! This problem has a very interesting history: Although originally posed by Steve Selvin in a letter to *The American Statistician* in 1975, it became famous when it was asked as a question to Marilyn vos Savant by a reader of her *Parade* magazine column "Ask Marilyn" in 1990. Ms. vos Savant, a very intelligent woman (actually listed in the *Guinness Book of World Records* from 1986 to 1989 as the person with the highest IQ) but herself not a mathematician, gave the correct answer, after which she received a storm of protesting letters from professional mathematicians and scientists — Ph.D.'s! — from all over the country. But Marilyn vos Savant was right, and the professionals who disagreed with her solution were all wrong, and so damn self-righteous about it! I don't know if she ever convinced them of the correct answer, but I have thought of a way of doing this, so here it is:

If you don't trade, then you will win just in case you originally chose the correct door, right? But if you do trade, then you will win just in case you originally chose the wrong door (because then the host will have opened up the other wrong door, and you will trade to the right one). Now, isn't it twice as likely to originally choose one of the two wrong doors than the single right door? Thus, if you do trade, you are twice as likely to win than if you don't.

R.: Suppose I make you the following offer: I give you two hundred dollar bills and make a statement. If the statement is false, then you must give me back one of the bills I gave you and keep the other, but if the statement is true, then you are to keep them both. Would you accept this offer? What are your reasons why?

{One respondent wrote that she would take the deal, because even if the statement proved false, she would still be $100 ahead. Another wrote, "Yes, I'd take the offer, for the same reason. But I am sure we are missing something essential. It can't be **that** easy. Or can it?"

I replied that it is NOT that easy! You should definitely NOT take my offer, because (I am proud to say) my offer is utterly diabolical!! If you took the offer I could make a statement such that the only way you could keep to the agreement is by paying me a thousand dollars (or any other amount that I could choose)!

[You didn't know that I was that evil, did you?] Anyway, I then changed the problem to the following: What statement could I make such that the only way you could keep your word and stick to the agreement is by paying me a thousand dollars?}

juufa72: What about "give me a thousand dollars or I will kill you." If you would say that, then there is a great chance that a person would give you the thousand dollars. I probably do not understand this puzzle.

R.: hahaha, juufa72. That's funny! There would indeed be a great chance that he would then give me a thousand dollars. But it is not logically necessary that he do this in order to adhere to the agreement I made with him. Remember, we would have agreed that if my statement were false, you would give me back one of the bills I had given you and keep the other, and if my statement were true, then you would keep both bills. That is the agreement. Now, I am claiming that I could make a sentence such that the only way you could adhere to the agreement would be by paying me a thousand dollars. What sentence would accomplish this? Compound sentences are allowed.

Alf: I've come up with a solution. A statement that works could be, "You'll neither keep my 2 hundred dollar bills nor give me a thousand bucks." If I remember well, this puzzle is a variation on the puzzle by which you won a kiss from your late wife.

R.: I think it is now a good time for us to conclude the discussion of the "diabolical" puzzle, which we recall is that I make you the offer that I give you 2 hundred dollar bills and make a statement. If the statement is false, then you must give me back one of the bills I gave you and keep the other, but if the statement is true, then you are to keep them both. What statement could I make such that if you adhere to the agreement, you have no choice but to pay me a thousand dollars?

There are several sentences that would work. The one proposed by Alf does indeed work, but the one I had in mind (which is perhaps a bit simpler) is this: "You will either give me back just one of the bills I gave you or give me a thousand dollars."

I will first show that this sentence cannot be false (assuming, of course, that you keep to the agreement). Well, suppose the statement were false. Then you would have to give me back just one of the bills, as agreed, but doing so would make the sentence TRUE (that you either give me just one of the bills, or a thousand dollars), and so we would have a contradiction! Therefore the sentence can't be false; it must be true. This means that you either give me just one of the bills or give me a thousand dollars. But you can't give me back either (or both) of the two $100 bills I gave you, because the agreement was that if the statement is true you are to keep both bills I gave you. Therefore you must give me a thousand dollars.

OK, here is a truly diabolical puzzle. Suppose I offer you a million dollars to answer a yes/no question truthfully. Would you accept the offer? If so, you shouldn't, because I could frame a question such that the only way you could answer truthfully is by paying me a BILLION dollars! Can you figure out what such a question could be?

pianolady: Raymond, it appears you have stumped us all on this one.

R.: OK! I offer you a million dollars to answer a yes/no question truthfully and you agree. But (oh horrors!) there is a question I could frame such that if you keep to the agreement, you have to pay me a billion dollars! There are several questions that would do the job, and one such is the following: "Will you either answer NO to this question or pay me a billion dollars?"

I am asking whether at least one of the following two alternatives holds:

(1) You will answer NO.
(2) You will pay me a billion dollars.

If you answer NO, then you are denying that either alternative holds, whereas the first alternative does then hold. Hence NO cannot be the correct answer. Hence to be truthful you must answer YES, and by doing so, you are affirming that at least one of those two alternatives does hold. But it can't be the first (since you didn't answer NO). Hence it must be the second, and so you owe me a billion dollars! [You see why this is called COERCIVE LOGIC!]

Now for another problem: There is a yes/no question that I could ask you such that it is impossible for you to answer it correctly. If you answer yes, you will be wrong, and if you answer no, you will also be wrong! Yet one of the two is the correct answer. It is just impossible for YOU to give it! The funny thing is that anyone in the world other than you could give a correct answer, but you cannot.

You are the only person in the whole wide world who cannot answer the question correctly with a YES or a NO. (By YOU I mean the person to whom the question is addressed.) What question would work?

I hope this problem intrigues you!

R.: Let us recall the problem: There is a yes/no question that I could ask you such that it is impossible for you to answer it correctly. If you answer yes, you will be wrong, and if you answer no, you will also be wrong! Yet one of the two is the correct answer.

The question I had in mind was: "Is your answer to this question 'no'?" If you answer YES, you are affirming that your answer is NO, which it isn't. On the other hand, if you answer NO, you are denying that NO is your answer, which in fact it is, so again you are wrong! Now, why could someone other than the person to whom the question is posed answer the question correctly with a yes or no? First you have to realize that if the name of the person to whom the question was posed was John, say, then the question from the point of view of any listener is "Is John's answer to the question 'no'?" Indeed, *after* John has responded to the question in whatever way he chooses (even by saying he's unwilling to answer at all, for instance), a listener (or anyone else who had been correctly informed about what John did) could answer correctly, for s/he would know whether or not John answered NO. For instance if John answered YES, or "bullocks!" or nothing at all, the listener could correctly answer NO. And if John answered NO, the listener could correctly answer YES. Moreover, *anyone* other than John could also answer the question without causing the paradox that John would get into if he answered either "yes" or "no", and could answer without problems even before John's response to the question! Such a person could, for instance, guess "yes" or "no" at random, and then, after John had done whatever he was to do, it could be determined whether the person had answered correctly or incorrectly. In other words John can't answer "yes" or "no" either truthfully or falsely because in both cases, we arrive at a contradiction. But anyone else *who knows what John did* can answer correctly. Moreover, anyone else can answer the question as soon as the question is asked (perhaps correctly, perhaps incorrectly) without any paradox arising, with what John actually did determining whether or not they answered correctly or incorrectly.

Here is a classical logic puzzle: You and two others are given a logic test. Each of you is blindfolded and a hat is put on your heads. You are told that each hat is either red or blue. The blindfolds are removed and each of you can see the other two hats. You see that the other two are blue.

You are then told, "If you see at least one blue hat, raise your hand." Of course you all raise your hands. Then one of the other two is asked, "Can you deduce the color of your hat?" He replies NO. What is the color of your hat, and why?

Here is the answer I would give if I were tested and this happened: Call the other two with me Arnold and Bernard, and assume that Arnold is the person who was asked whether he could deduce the color of his hat. If my hat were red, then Arnold, having seen Bernard raise his hand, would know that Arnold had seen a blue hat. And since mine is red, Arnold would know that it was his blue hat that Bernard saw. But Arnold had said he couldn't deduce what color his hat was, so that my hat can't be red.

I heard the above classic problem of the three hats when I was in high school, and I then thought of a complex variant of it which I subsequently published in the mathematical games column of the *Scientific American*. Here is my variant (which is a GOOD DEAL MORE COMPLEX!).

Three men — A, B, and C — were aware that all three of them were PERFECT LOGICIANS in the sense that each could instantly deduce all consequences of a given set of premises. They were shown four red and four blue stamps. While they were blindfolded, two stamps were pasted on each man's forehead, and the remaining two stamps were put away in a drawer. When the blindfolds were removed, each one could see the stamps on the foreheads of the other two. Actually each one had one red stamp and one blue stamp on his forehead. First A was asked if he knew what he had, and he answered NO. Then B was asked if he knew what he had, and B answered NO. Then C was asked the same question, and he answered NO. Then A was asked if he now knew what he had and A answered NO. Then B was asked if he now knew what he had, and B answered YES. How did B know what he had?

I realize that this problem is far from easy, but even if you don't solve it, I think the solution will interest you.

pianolady: I'm going to wait for the solution. The last one wore me out.

R.: That was a difficult puzzle! Here is the solution:

By the time his turn arrives in the second round, *B can reason as follows*: Suppose my stamps were of the same color — say both red. Then at the beginning of the game A would have immediately seen three reds — my two and C's one — and know that he couldn't have two reds. By the time his turn came during the second round, he (A) would also know that he couldn't have two blues (so must have a blue and a red), because he is a good logician and would reason that if he had two blues then C would have immediately seen (i.e. at the beginning of the first round of the game) two reds and two blues, and hence would have known at his turn in the first round what he (C) had on his forehead, because he would reason as follows:

> If I have two reds, then A would have immediately seen four reds, and would have known on the first round that he has two blues, which he didn't. Therefore I can't have two reds.

> If I have two blues, then B would have immediately seen four blues, and would have known on the first round that he had two reds, which he didn't. Therefore I can't have two blues.

> Hence I have a red and a blue.

Thus, B knows, when his turn comes in the second round, that if he had two reds, then A would have known what he (A) had when his (A's) turn came in the second round, which A didn't. So B knows at his turn in the second round that he cannot have two reds. By a symmetric argument, B knows at his turn in the second round that he cannot have two blues. Thus B concludes when it is his turn in the second round that he must have a red stamp and a blue stamp on his forehead.

R.: I just saw in the audition room some discussion of the Beethoven Hammerclavier Sonata, and some stories come to mind.

(1) A man walked into a bar with a mouse, a bird and a tiny piano. First the mouse sat down at the piano and played the entire Hammerclavier Sonata. Then the bird sang the entire Schubert Wintereiser accompanied by the mouse at the piano. The owner was amazed and

offered the man $50,000 to buy the act. At first the man agreed, but then his conscience bothered him and he said, "No, no, no. I can't cheat you like that. The whole thing is a fake! That bird never sang! The mouse is a ventriloquist!"

(2) It is rumored that no one can play the last movement of the Hammerclavier Sonata at sufficient speed. Well, in the science fiction novel *The Black Cloud* by Fred Hoyle, an advanced alien in the form of a gigantic black cloud was approaching the sun in such a way that the end of all life on Earth was clearly going to be the inevitable consequence of its stay in the solar system. Already millions of lives were being lost. Scientists finally managed to establish radio contact and teach the alien how to communicate with them in English. After having learned much about life on Earth, the alien asked the scientists to send a sample of human music. The scientists sent a performance of the Hammerclavier Sonata to the alien. Almost immediately the message came back: "It should be faster!"

(3) I once told this story to a pianist who then told me of the following true incident: He was sitting on a bus behind two ladies, and he could hear their conversation. One asked the other whether she liked classical music. The other replied, "No, I like things that are faster."

techneut: I love the one about the mouse and the bird.

abimopectore: Wow, I really should visit this forum every day. It looks like I missed a lot …!

R.: Here is a little puzzle I just thought of: You have two boxes labeled Box 1 and Box 2. Each box is either empty or contains a prize. It could be that both are empty or both contain prizes, or that just one of them contains a prize. A sentence is written on the lid of each box as follows:

On Box 1: "This box contains a prize and the other box is empty."

On Box 2: "One of these boxes contains a prize and the other one is empty."

It is given that one of the sentences is true and the other is false. Does Box 1 contain a prize? What about Box 2?

techneut: If the sentence on Box 1 were true, then so would the one on Box 2. But that is not possible, as one of them must be false. So the sentence on Box 1 must be false, and hence the one on Box 2 is true. So one and only one of the boxes has a prize in it. This must then be Box 2, for otherwise the statement on Box 1 would be true, which it isn't.

So Box 1 is empty and Box 2 has the prize.

R.: Good work, techneut!

Now comes a series of puzzles I will be making up especially suited to the Piano Society! Here goes the first: In some far off ocean there is a very, very, VERY strange island named MUSICA in which every inhabitant is either a pianist or a violinist, but no inhabitant is both. The amazingly curious thing about this island is that every lady pianist always tells the truth and every lady violinist always lies, whereas the men are the opposite — the male pianists lie and the male violinists tell the truth! [How do you like this setup?] Well, on one of my visits to this weird place, I came across two inhabitants, L (a lady) and M (a man), who made the following statements:

> L: We are both pianists.
> M: That is true.

Is the lady a pianist or a violinist? And what about the man?

techneut: Suppose L spoke the truth. Then, because of that she is a pianist, and because her statement is true, M is also a pianist. But M, being a male pianist, could not have said that L's statement is true. So the assumption leads to a contradiction.

Consequently, L must have lied and is therefore a violinist. Now M says her statement is true, which we know it isn't, so he is lying and therefore he's a pianist.

I would like to hear the two of them rehearse the Brahms Sonatas.

R.: Here is another puzzle about this very curious place: On my next visit there, I met another lady and man whom we will again respectively call L and M, who made the following statements:

> L: We both play the same instrument.
> M: She is a pianist.

What instrument does each play?

NOTE TO TECHNEUT: Techneut, you are obviously excellent at these logic puzzles, and so I suggest that when you solve them, you don't immediately post a solution, but simply say something like "I got it!" and thus give others a chance to post their solutions. Then, if you like, send me your solution by a PM.

pianolady: Raymond, it's OK. Techneut doesn't have to post his answers privately. There is no fun in that. What about this idea? Put up your new posts in the morning time (American time). I don't know when you normally wake up, but if you post at 8:00 am your time, then it's 7:00 am my time and either 1:00 or 2:00 pm (can't remember) in Europe. Or give and take an hour or two each way. This way, everybody has the chance to view the puzzle when it's daytime, instead of seven or eight hours later as what happens if you post late at night (our time) and some of us are going to sleep and the others are just waking up.

R.: Here is the solution to the last puzzle about Musica. From just L's statement alone, it follows that M must be a pianist. Because suppose that L is a pianist. Then she is truthful, and hence he is a pianist like her. On the other hand, suppose she is a violinist. Then she lied, and so they play different instruments, which means that he is again a pianist. Since he is a male pianist, he lied, which means that she is a violinist — and also lied. [Quite a pair!]

Here is a rather easy puzzle about the strange island of Musica: On one of my visits there, I went to the home of a Mr. and Mrs. Smith. They owned a nice-looking piano that I saw in their living room. I asked the wife whether it was a Steinway. She replied, "I am a violinist and our piano is not a Steinway."

Was the piano a Steinway or not?

R.: Pianolady, you are right that the piano is a Steinway. The clearest way to see it is to first note that the wife couldn't be a pianist (for reasons that you correctly gave), and so she is a violinist. Now if it were true that the piano was not a Steinway, then it would be true what she said, namely that she is a violinist and the piano is not a Steinway. But lady violinists don't make true statements. Therefore the piano must be a Steinway.

On another of my journeys to Musica, I visited the home of a married couple. I sat on the sofa between the husband and wife. At some point the one on my left said, "My spouse is truthful." The other one then said, "My spouse is a pianist."

Who sat on my left, the wife or the husband?

R.: Solution To The Last Problem: Let A be the one to my left and B the one to my right. Since A said that B is truthful, then the two are either both truthful or lying (because if A is truthful, then B is also truthful, as A truthfully said, and if A is lying, then B lies, contrary to what A falsely said). First consider the case that both are truthful. Then, as B truthfully said, A is a pianist. Thus A is a truthful pianist, and hence a lady. Now consider the case that both lied. Since B lied, then A must really be a violinist, and being a lying violinist, is again a lady. Thus in either case, A is a lady, and so it is the wife who sat on my left. There is no way of knowing whether she is a pianist or a violinist.

A MUSICAL QUATRO: Suppose you receive an email statement from a member of the Island of Musica. Here are 4 questions:

(1) What statement could convince you that the sender must be a lady violinist?
(2) What statement could convince you that the sender must be a lady pianist?
(3) What statement could convince you that the sender was a lady, but you had no way of knowing what instrument she plays?

(4) What statement could convince you that the sender was a pianist, but you had no way of knowing whether the sender was male or female?

ANSWERS TO THE MUSICAL QUATTRO

(1) I am a male pianist.
(2) I am not a male violinist.
(3) I am a pianist.
(4) I am female.

EXPLANATIONS (in case you need them)

(1) No truthful person would claim to be a male pianist (who lies). Hence the statement was false, which means that the sender was not really a male pianist, but being a liar, must therefore be a female violinist.
(2) If the statement were false, that would mean that the person WAS a male violinist, but male violinists don't make false statements. Hence the statement must be true. Thus it is true that the person is not a male violinist, but being truthful must therefore be a female pianist.
(3) A lady pianist could truthfully say that, and a lady violinist could lie and say that. But a male violinist, being truthful, would never say that, and since a male pianist lies, he would never make the true statement that he is a pianist. Thus any lady could say that, but no man could.
(4) A female pianist could truthfully say that, and a male pianist could falsely say that, but a female violinist could not make the true statement that she is female, and a male violinist would never lie and claim to be female. Thus any pianist could say that, but no violinist could.

There is a pretty symmetry between (3) and (4). Only a female could claim to be a pianist, and only a pianist could claim to be female.

I must admit I am getting quite fond of this island of Musica, and yesterday I thought of three more questions about this island:

(5) What statement could be made by anyone other than a lady violinist?

(6) What statement could be made by either a female pianist or a male pianist or a female violinist or a male violinist? Any member of the island could make that statement!

(7) I got an email from a member of Musica from which I could deduce that the sender must be either a female pianist or a male violinist, but I had no way to tell which! Moreover, the message was only two words long! Can you supply such a statement?

respondent: Raymond, just to let you know, I'm sort of tied up these days and can't find the time to do this puzzle.

R.: Answers to my last puzzles:

(5) I am not a male pianist.

(6) I am truthful.

(7) I exist. [This statement is obviously true, and only female pianists and male violinists make true statements.]

Number 7 reminds me of two jokes about the philosopher Rene Descartes who begins his whole system with the famous words, "I THINK, THEREFORE I EXIST." The story is told that Descartes was at a bar and the bartender asked him, "Monsieur Descartes, would you like a cocktail?" Descartes replied, "I think not" and disappeared.

Another one: A man (*M*) goes into a bar and has the following conversation with the bartender (*B*):

M: You don't know who I am, do you?

B: No.

M: You have never seen me before, have you?

B: No.

M: But you just saw me walk in the door?

B: Yes.

M: Then how do you know it's me?

There is the true story that the philosopher and psychologist William James once asked a boy if he knew what faith is. The kid replied, "Yeah! Faith means believing something you know ain't true!"

Speaking of faith, the following is very puzzling to me: On television I constantly hear of a conflict between evolution and so-called intelligent design. This is a completely false dichotomy! There is no conflict between evolution and intelligent design! The real conflict is between evolution and CREATIONISM, which holds that humans did not evolve from lower life forms. It seems that people are using the phrase "intelligent design" to mean creationism, but this should not be done, as it only creates confusion. As Confucius wisely said, "If I become governor, the first thing I would do is rectify names."

Terez: The problem is "Intelligent Design" is propaganda used by creationists who are trying to get it taught in schools as an alternative to evolution, so it's a bit difficult to separate the two. Evolution disputes the Biblical account of creation, whether or not it disputes a (much) less literal interpretation of that account. The creationists in question are using "Intelligent Design" to represent the Biblical account of creation.

techneut: **rsmullyan wrote:** Many people who believe in evolution believe that evolution itself was intelligently designed!

If evolution can be considered a design at all, it does not seem very intelligent to me … . It boils down to randomly trying out zillions of possibilities and let time decide which work and which don't. Then again, had evolution been more intelligent, we humans would probably have either exhausted or destroyed the planet by now …

R.: In my book, *Who Knows? A Study of Religious Consciousness* (Indiana Press), I raise several questions, two of which I will tell you in this and my next posting. The first is addressed to those who believe in an afterlife and in Heaven and Hell — particularly to those who believe in the reality of Hell. There has been much controversy as to the moral justification of Hell. The Russian Christian philosopher Nicolai Berdyaev wrote, "In the past, the idea of Hell has driven many into the church. Now, the idea has driven many out of the church. If there really is a hell, then I am an atheist." Similarly, Lewis Carol wrote a letter to his sister in which he said, "If I believed that Christianity really implied the existence of Hell, then I would give up Christianity." On the other hand, there are some ministers who have said that the idea of Hell seems horrible to the unenlightened, but the enlightened will realize that the idea of Hell is reasonable and has scriptural support. And so you see that this topic has engendered much controversy. Now, my question to those who believe in the reality of Hell is this: Suppose that when you get to Heaven, to your great surprise, God says to you, and all the other saved ones, "I know there has been much controversy as to whether Hell is morally justified or not. Of course I have my own ideas on the subject, but I want you all to be happy, and so I am going to let you vote on the matter — vote on whether Hell should be abolished or retained. If more than half of you vote on its abolition, then I will abolish it. Otherwise I will retain it."

My question is: How would you vote? I got some interesting replies to the question, which I will tell you about in my next posting, but for now, I am curious to know how those of you who believe in the reality of Hell (if there are any) would answer the question. Comments from others would also be of great interest. In my next posting I will also tell you my motive in raising the question.

R.: Concerning my question about Hell, most people I have asked have said that they would vote for the abolition. One person said that the choice would be very difficult. Another person told me that he would vote for the retention on the grounds that "Heaven and Hell are two sides of the same coin — you can't have one without the other." The cleverest answer I got was from a Catholic theology student at the University of Notre Dame. He said, "I would vote for the abolition, but that may well be an imperfection on my part."

My purpose in asking the question was to determine whether those who believed in and didn't disapprove of Hell did so because they really thought it was right, or whether they did so because they believed their religion REQUIRED them to do so!

A second question I raise in my book is how the reader would like the following scheme, which I call "collective salvation" — namely, on the Day of Judgment, God takes the average of all the past deeds of the entire human race, and we all either rise or fall. I asked various people how they would like that idea. One person responded, "Oh, that's a dangerous idea!" To my great amusement, another responded, "I wouldn't like that at all! I think my chances would be much less!"

beclever: I believe that good people would be good without the concept of Hell, but it sure feels better to think that there is some final justice for the evil-doers in the world (whether they are minor evil-doers like line-cutters or major evil-doers like Enron executives).

That's one of the hardest things about being agnostic: you have to be ready to accept the fact that jerks can be jerks and get away with it. I guess that's what makes the choice about keeping heaven but abolishing hell so hard. Can you live with the fact that bad people can live badly and still go to heaven?

R.: Many people believe that what is called for is not punishment but reformation. Eastern religions such as Hinduism and Buddhism believe that ultimately all souls will become enlightened. Also, as one Universalist Unitarian minister wrote, "I believe that ultimately existence will be a blessing to everyone." As another put it, "It seems inconceivable to me that a good God would ever create a being who through misuse of his own free will, or for any other reason, would land himself in a situation in which he would suffer eternally." One person I know once said, "Could you really be happy in Heaven, knowing that some are suffering eternally in Hell?"

Terez: Well, according to the most common Christian philosophy, you don't even have to be a bad person to go to hell … just a non-Christian. "Being good" doesn't get you to heaven — only your acceptance of Jesus as your Savior can get you to heaven (whatever that means ... there are as many interpretations as there are Christians).

R.: I am not sure that what Terez said is really true for MOST Christian sects. It is certainly true for many Protestant sects. The Catholic Church doesn't hold that only Christians can be saved.

Before leaving the subject, I should mention the Swedenborgian notion of Hell, which many regard as the most humane of all: According to Swedenborg, after death, the person goes directly to Heaven, but the evil people cannot stand the presence of goodness, so they voluntarily depart to another place filled by equally evil people. They are all so wicked and selfish that they are constantly hurting each other terribly. God sends angels there to relieve as much suffering as possible. Thus the Swedenborgian model of Hell is not like a penal institution, but more like an insane asylum!

Terez: The Catholic Church has changed a lot since Vatican II ... every Catholic I know has the same "Jesus saves" philosophy as the Protestants. Perhaps it doesn't mesh well with the Purgatory idea, but there it is. Mormons believe that people will be given an opportunity to repent after death.

techneut: The Biblical divisions between heaven/hell, good/bad, light/dark etc. are simplistic, archaic and ridiculous. People can't be divided into good ones and evil ones. There are infinitely many gradations, as in everything. I guess it will be a standard bell curve. Nobody is totally evil, and nobody is totally good. Most people will be around the median, combining good and evil in comparable measures. That evil doesn't come out as easily as in past ages is largely due to civilization and laws, but that doesn't mean it is not latent in many people. Is one going to hell for evil done or evil thought (perhaps even subconsciously)? It is a difficult subject best left alone ...

R.: Terez, I believe that Muslims also believe that one can repent after death. As I understand it, once one is in Hell, if one begs Allah for forgiveness, he will forgive.

techneut: Muslim hell will be an empty place then! Most of them will take that option I guess.

R.: Now for a puzzle. Suppose I would make you the following offer: I hold in my hands a ten-dollar bill and a hundred-dollar bill. You are to make a statement. If the statement is true, then I promise to give you one of the bills, not saying which one, but if the statement is false, then I give you neither bill. What statement could you make which would force me to give you the hundred-dollar bill (assuming, of course, that I kept my word)?

beclever: Raymond, when you say "force", do you mean "force" in the purely logical sense (i.e. 1 ~= 0), or "force" in the sense that you are human and therefore presumably susceptible to financial inducements and/or threats of physical maltreatment?

R.: Beclever, all I meant is that there is a statement you could make such that the only way I could keep my word is to give you the hundred-dollar bill.

R.: Beclever, you are remarkably close, but there is a subtle but significant way of improving your formulation that I will tell you about shortly. My solution, although very close to yours, is slightly, but significantly, different, and is that you say, "You will not give me the $10 bill." The only way the statement could be false is that I do give you the $10 bill, but I am not to give you either bill for a false statement, hence the statement can't be false. It must be true, which means that I will not give you the $10 bill. Yet I must give you one of the two bills for a true statement, and so I have no choice but to give you the $100 bill.

diminished second: Hahaha. My piano teacher is always coming up with horrible puns. One of them goes as follows.

There was an artist named Mo. Mo had some of his art on display at a local museum, along with a few other artists' works. When the display was taken down, they were supposed to leave Mo's work up, and take down all the rest. However, there was a mistake in communication, and all of the art was taken down. Later, the manager came in and said, "Where are the pieces that Mo painted? The guy taking down the display was supposed to leave them up!" They looked around, and finally one of the guys found Mo's stuff packed away in a box with all the rest of the art. When he found this, he said, "Box hidin' Mo's art!" ("Bach's" "Haydn" "Mozart" in case you didn't get the triple pun there. 😊)

pianolady: On behalf of Raymond ... I recently came across two rather cute jokes.

(1) Why is it difficult for mummies to make friends?
    Answer: They are too wrapped up in themselves.

(2) A man told a friend that he had the largest sheep farm in the state. When asked how many sheep he had, he replied, "I don't know. Every time I start counting them, I fall asleep."

R.: Here is a nice controversial puzzle. Three men — Arthur, Bernard, and Charles — were about to travel through the Sahara desert. They stopped one night at an inn at the outskirts of the desert. Arthur hated Bernard and wanted to murder him, so he put poison in the water of Bernard's canteen, which would be the only source of Bernard's water. Charles also wanted to kill Bernard, and knowing nothing about what Arthur had done, drilled a tiny hole in Bernard's canteen so that all the water would spill out. As a result, Bernard died on the trip from lack of water. The question now is, who was responsible for Bernard's death, Arthur or Charles? What do you think?

[Answer from a later post:] As I said, this is a controversial puzzle, and so it is questionable whether there is just one correct answer. My own opinion is that Arthur was the murderer, because once Arthur acted, Bernard was doomed. I cannot see how removing poisoned water from Bernard's canteen could hurt Bernard in the least. But again, this is but one man's opinion. I would be curious to know what would happen if this case really came up and Arthur and Charles were tried in court. Of course both were guilty of attempted murder, but which one would be convicted of actual murder?

Now here is a very puzzling situation: A man was tried for murder and the jury found him guilty of murder in the first degree. The judge said to the defendant, "This is the strangest case I have ever handled. Not only can I not sentence you to death, I can't even sentence you to jail!"

How is this possible?

R.: I'm talking about a situation that could really happen! The solution is that the defendant was a Siamese twin. The other twin was innocent (perhaps asleep at the time). One cannot execute, or even incarcerate, an innocent person.

Now for a riddle: You know what a dock is (a boat usually has a place to dock). Why is it logically impossible for there to be more than one dock in the universe?

juufa72: Because you would have a "paradox" (pair of docks).

R.: Juufa72 is right. The reason that it is logically impossible for there to be more than one dock in the universe is that if you have so much as two, you will have a pair-a-dox.

Here is a special puzzle: A reporter visits a certain community and publishes the following account. The inhabitants of the community have formed various clubs. Each club is named after one and only one inhabitant and each inhabitant has one and only one club named after him. An inhabitant could be a member of more than one club. It is not necessary that a person be a member of the club named after him. If he is, then he is called SOCIABLE. If he isn't, then he is called UNSOCIABLE. It so happens that the set of unsociable inhabitants forms a club.

The question now is this: Is the reporter's account consistent or not?

beclever: The situation described cannot exist. Proof: Since the group of UNSOCIABLE people form a club, by the rules of naming the clubs, the club has a name that is the same as the name of an individual in the community. Let's call this individual X (and therefore the name of the club is X). Is X in this club or not?

If X is in the club, by definition of the club, X is UNSOCIABLE. But because he is a member of the club that bears his name, he is SOCIABLE, which membership implies he is not a member of the club. A contradiction.

If X is not in the club, then since the club's name is X, X by definition is UNSOCIABLE, and therefore X is a member of the club. Contradiction. QED.

R.: Juufa72 correctly stated that the situation is impossible, but did not prove it; beclever stated that the situation is impossible and correctly proved it. It is not necessary for me to repeat his correct proof.

IS THERE A SPY IN THE COMMUNITY?

We now visit another community, which is like the last one in that the inhabitants have formed various clubs and each club is named after one and only one inhabitant, and the name of each inhabitant X is the name of one and only one club, called X's club. Unlike the last community, an inhabitant who is a member of a club can be so either openly or secretly. And inhabitant X is called AMICABLE if X is openly a member of X's own club. To say that a person X is not amicable is to say that X is either not a member of X's club at all, or else is a member, but secretly so. If the latter — if X is secretly a member of X's club — then X is called a SPY. Whether there are any spies in the community has been a highly controversial topic in this community for many generations. According to one view, it is impossible for there to be a spy there. According to another view, it is certain that the community must contain a spy. According to a third view, it is impossible that there is one, and it is possible that there isn't one. The community finally decided to call in a logic detective to see if he could throw any light on the matter. After much research, the detective uncovered the following two facts:

> Fact 1. For any club C, the set of all inhabitants who are not members of C forms a club of its own. [Incidentally, in mathematics, this club is denoted C', and is called the COMPLEMENT of the set C.]

> Fact 2. The set of all amicable inhabitants is a club.

From these two facts, the logic detective could determine whether there was a spy in the community or not? Was there? Why?

beclever: In Raymond's new town, there is at least one SPY. To see this, note that the set of all AMICABLE people form a club (fact 2) and so do the set of all UNAMICABLE people (fact 1).

Since the set of UNAMICABLE people form a club, this club has a name which is the same as an individual in the community. Let's call this person GWB (and hence the club's name is GWB).

If GWB is AMICABLE, then he is openly a member of the club which bears his name. But by definition this club is the same as the set of UNAMICABLE people in the town, which means that GWB is UNAMICABLE. Hence GWB cannot be AMICABLE, and therefore must be UNAMICABLE.

UNAMICABLE means an individual is either not a member of the club that bears his name or is a secret member of that club. Since GWB must be a member of the club GWB (by the previous paragraph), he must therefore be a secret member and therefore a SPY.

R.: beclever was correct in his proof that there must be a spy in the community.

PROBLEM OF THE LISTED SETS

We are given a very curious book — curious because it contains infinitely many pages — Page 1, Page 2, ..., Page *n*. For each positive whole number *n* there is a Page *n*. On each page is listed a description of a set of positive whole numbers. Is it possible that EVERY set of positive whole numbers is listed in the book, or is it necessary that there be at least one set of positive whole numbers that cannot be listed anywhere in the book?

The answer to this question is amazingly simple, yet of profound mathematical importance, for work on this question and related ones started the whole field known as set theory! Those questions were first posed, and the answers and their proofs discovered, by the 19th century mathematician Georg Cantor. The statement of the solution to the above question is known as Cantor's Theorem. Despite the simplicity of the question and of the proof of its solution (once seen), it took thousands of centuries before it was posed and discovered!

For you to solve this great problem, here is a hint: Recall the problem of the community with the clubs and the proof that the set of unamicable people cannot be one of the clubs. Well, the present problem is very closely related to that problem. In fact it can be regarded as a special case of it.

p.s. Beclever already knows the solution.

To say more, think of the numbers as people, and think of the listed sets as clubs, and think of each number *n* as the name of the set listed on Page *n*. Now recall the club problem and the proof that the set of unsociable inhabitants cannot be one of the clubs.

Alright, call a number EXTRAORDINARY if it is a member of the set described on Page *n*, otherwise ORDINARY. What about the set of all extraordinary numbers? Could that set be listed on any page? What about the set of ordinary numbers?

smulioni: I'll wager that the set of all extraordinary numbers, by the above definition, could appear in the book (although I believe it could be shown that it doesn't need to). If the page on which the description, "the set of all extraordinary numbers", appears is E, E could be either an ordinary number, which does not belong to the set described on Page E, or an extraordinary number, which does. The set of all ordinary numbers, on the other hand …

R.: Smulioni is right that there is no way to know whether the set of extraordinary numbers is or is not listed on some page of the book, but the set of ordinary numbers cannot be. Someone else, please say why!

R.: Alright, here is a proof that the set of all ordinary numbers cannot be described on any page of the book. Let Set 1 be the set described on Page 1, Set 2 the set described on Page 2, and so forth. Thus for each number *n*, Set *n* is the set described on Page *n*. If *n* belongs to the Set *n*, then *n* is called extraordinary, and if it doesn't, then *n* is called ordinary. Now suppose the set of all ordinary numbers was described on some page of the book — say Page 13. Then we would have the following contradiction: Set 13 is now the set of all ordinary numbers, which means that every ordinary number is in Set 13, and no extraordinary number is. What about the number 13 itself — is it ordinary or not? Either way we get a contradiction! To say that 13 is extraordinary is to say that 13 is in Set 13, but that can't be, since only ordinary numbers are in Set 13. On the other hand, if 13 is ordinary, it must be in the set of ALL ordinary numbers, which is the very set 13. Thus if 13 is ordinary, then 13 belongs to Set 13, which makes 13 extraordinary, and we again have a contradiction! Of course the same argument works for any number *n* other than 13, and so the set of all ordinary numbers cannot be described on any page of the book.

Here is a riddle for you: If the tail of a dog would be called a leg, how many legs would a dog then have?

pianolady: My guess is the dog would still have four, because a leg is something you stand on. I don't think I've ever seen a dog stand on its tail. ☺

And that reminds me of a dumb joke, which you all probably know. What do you call a woman with only one leg?

beclever: I was thinking she would be called Peg.

pianolady: Oh, that's good too! ☺

R.: Pianolady was right about the riddle, "If the tail of a dog was called a leg, how many legs would a dog then have?" The answer is four, because calling a tail a leg doesn't mean that it is one.

The riddle was originally posed by Abraham Lincoln, and it had political significance. Someone once said to him that a certain thing which clearly was slavery wasn't slavery, and Lincoln then gave him this riddle to illustrate that saying that something is so doesn't mean that it really is so.

Let me now tell you this story about Abraham Lincoln: An author tried to sell one of his books to President Lincoln. Lincoln was not interested. The author then said to him, "Well, since you are not interested in a copy for yourself, would you at least write an endorsement for it to make it easier for me to sell it to others?" Lincoln very kindly said, "Certainly," and then wrote the following endorsement:

THOSE WHO LIKE THIS KIND OF BOOK
WILL FIND IT JUST THE KIND OF BOOK
THEY LIKE!

R.: A violinist of a certain orchestra always had a pained expression on his face whenever he was playing. One day the conductor had a private conversation with him:

Conductor: What is bothering you? Don't you like the way I conduct?
Violinist: It's OK.
Conductor: Then don't you like the programs I choose?
Violinist: They are alright.
Conductor: Maybe you don't like the way the others play?
Violinist: They are OK.
Conductor: Then what is bothering you?
Violinist: I don't like music!

R.: I like the following definition of a specialist: A specialist is one who does everything else worse.

I recently heard the following definition of a fugue: A fugue is that musical form in which one voice after another comes in and one listener after another goes out.

pianolady: That's perfect! 😊

R.: A man was testing his wife's hearing. While her back was turned he said, "Can you hear me?" No answer. He stepped closer and said, "Can you hear me?" No answer. He got still closer and said, "Can you hear me?" No answer. He shouted in her ear, "CAN YOU HEAR ME?" She replied, "I already said yes three times!"

I will be out of town for a few days. Again, our kind, lovely, charming, beautiful and talented pianolady will hopefully post my entries for me.

My two definitions of an oxymoron:

> Definition 1. A stupid ox.
> Definition 2. A moron educated at Oxford.

pianolady (posting on Raymond's behalf): About 70 years ago I heard the following joke: A man took a Pullman train from New York to San Francisco. Before he got off in San Francisco he asked the porter how much was his average tip. The porter replied that it was two dollars. The man gave him two dollars, at which the porter said, "You know, you are the first one to come up to my average!"

R.: Speaking of tips, I am reminded of the gag of Groucho Marx aboard ship. He asked the steward, "Is tipping allowed on this boat?" The steward happily said, "Yes, Sir!" Then Groucho asked, "Do you have change of ten dollars?" The steward replied, "Yes, Sir!" Upon which Groucho then said, "In that case, you won't need the nickel I was going to give you."

I once read the following story about the philosopher Moses Mendelssohn, the grandfather of Felix: He was evidently on good terms with Frederick the Great. One day while he was out walking, he met Frederick, who asked him where he was going. Moses replied, "I don't know!" The king was furious and said, "How dare you give me such an answer! Guard, take him to prison!" Moses replied, "You see, Majesty, did I know that I was going to prison?" This amused the king, who smiled and forgave him.

Once Houdini was sent abroad to tame a savage tribe. He did the following trick, which made all the natives certain that he had magical powers: In his hut he had on the floor a large wooden trunk with a handle on it. The strongest tribesmen couldn't lift it, but when Houdini said the magic words, the smallest boy could lift it!

Can you guess how it was done?

R.: The answer is that the bottom of the trunk was made of sheet metal, and underneath the floor was a powerful electromagnet that Houdini could control.

And now for some humor: Two cows were standing in a field. One said to the other, "With all this talk of mad cow disease, aren't you worried?" The other replied, "Why should I worry? I'm a helicopter."

Once, Melvin Fitting told me that the Chinese sages used to wear sandals, and that's why they were called toe-ists. I countered by saying that the Indian sages used to wear very high shoes, and that's why they were called boot-ists.

Let me now tell you one about Samuel Johnson. In a letter to an author, Johnson wrote, "Your manuscript is both good and original. Unfortunately the good parts are not original, and the original parts are not good."

There is also the true story that Disraeli wrote to an author who sent him a manuscript: "I can assure you, Sir, that I will lose no time in reading it!"

Another incident about Disraeli: Once in a parliamentary debate with Gladstone, the latter got very angry and said to Disraeli, "I predict, Sir, that you will die either by hanging or of some vile disease." Disraeli replied, "That all depends, sir, upon whether I embrace your principles or your mistress."

When I was 10 years of age, my brother, 10 years older than me, gave me the following definition of a gentleman: "A gentleman is one who does not hurt other people's feelings unintentionally."

One Christmas day in the old communist Russia, Rudolph (an ardent red) and his wife were looking out the window. The wife said it was snowing. Rudolph insisted it was raining. His wife asked him how he was so sure. He replied, "Because Rudolph-the red-knows-rain,-dear!"

I recently heard the following true story: An American tourist in Tokyo was looking for a bank. He came across a building in front of which were standing several Japanese. He asked them if they spoke English, which they did. He then asked them how to get to the bank. They then spoke excitedly among themselves in Japanese, which the tourist did not understand. Along came a friend of the tourist who understood Japanese and explained what they were saying: The tourist was standing in front of the bank the whole time, and the Japanese were trying to find a way of telling him without embarrassing him! How beautifully Japanese!

R.: Some critic said of Franz Liszt, "He has gone far, but has not advanced."

pianolady: I bet it was Clara Schumann who said that.

R.: No, it was some critic.

I love the story of the cellist Piatigorsky, who at the age of eighteen attached a cord to the cello of one of the older, dignified players in the Warsaw Orchestra, having previously arranged for a friend backstage to hoist the instrument gently into the air just as the bow was about to touch it during the performance. He never regretted the week's salary the little joke cost him.

Did you know that the Bolivian government was so proud of Jaime Laredo that they issued a stamp on which appeared La-Re-Do?

A music critic once actually wrote about the recital of a pianist, "His intonation was perfect!"

A salesman was once trying to sell a certain device to a customer by pointing out the good features, and at one point he said, "Moreover, this sells for less than those costing twice as much!"

A man once came across a girl reading a book on love-making. She turned to him and said, "It says here that the best lovers are American Indians and Poles. Did you know that? My name is Mary. What's yours?" He replied, "Red River Kovolski."

R.: Pianolady, you ask if I have any more jokes. Yes, here is one. A man said to his friend that his grandfather knew the exact year, date and time that he would die. When asked how the grandfather knew, the man replied, "The judge told him."

pianolady: Thanks, Raymond. I needed that! ☺

R.: Jack decided to go skiing with his buddy, Bob. So they loaded up Jack's minivan and headed north. After driving for a few hours, they got caught in a terrible blizzard. They pulled into a nearby farm and asked the attractive lady who answered the door if they could spend the night. "I realize it's terrible weather out there and I have this huge house all to myself, but I'm recently widowed," she explained. "I'm afraid the neighbors will talk if I let you stay in my house."

"Don't worry," Jack said. "We'll be happy to sleep in the barn. And if the weather breaks, we'll be gone at first light." The lady agreed, and the two men found their way to the barn and settled in for the night. Come morning, the weather had cleared, and they got on their way. They enjoyed a great weekend of skiing. About nine months later, Jack got an unexpected letter from an attorney. It took him a few minutes to figure it out, but he finally determined that it was from the attorney of that attractive widow he had met on the ski weekend. He dropped in on his friend Bob and asked, "Bob, do you remember that good-looking widow from the farm we stayed at on our ski holiday up north about nine months ago?"

"Yes, I do," said Bob.

"Did you happen to get up in the middle of the night, go up to the house and pay her a visit?"

"Well, yes," Bob said, a little embarrassed about being found out. "I have to admit that I did."

"And did you happen to give her my name instead of telling her your name?"

Bob's face turned red, and he said, "Yeah, look, I'm sorry. I'm afraid I did. Why do you ask?"

"She just died and left me everything."

pianolady: That's funny, Raymond. I've got one for you now:

> In a dark and gloomy room, the fortune-teller was startled by what she saw in her crystal ball. She looked at her customer sitting across the table and said, "There's no easy way for me to tell you this, so I'll just be blunt. Prepare yourself to be a widow. Your husband will die a violent and horrible death this year."

> Visibly shaken, the woman stared at the psychic's lined face, then at the single flickering candle, then down at her hands. She took a few deep breaths to compose herself. She simply had to know. She met the fortune-teller's gaze, steadied her voice, and asked, "Will I get away with it?"

R.: First for a joke: A man went to a priest and told him, "I am 90 years old and yesterday made love to three women." The priest said, "Really! When did you last go to confession?" The man replied, "I don't go to confession. I'm not Catholic." The priest said, "Then why are you telling this to me?" The man replied, "I'm telling this to everybody!"

This joke came to mind because today {this was posted in December, 2008} I received an advanced copy of my latest (my 22nd) book: *Logical Labyrinths* (AK Peters, Ltd.), which received two wonderful reviews on the back cover, and being the incurably immodest man that I am, I want to tell this to EVERYBODY! One of the reviews is from the famous mathematical logician Dana Scott (a member of the National Academy of Sciences):

> "This book overflows with wit and wisdom from a master of puzzles and paradoxes. I wish all my students had worked through this text. Suitable for self-study, the book will be most valuable if the reader agrees not to peek at the solution before learning to think for themselves."

The other is from Martin Gardner (one of the world's most, if not THE most, famous experts on mathematical games and puzzles):

> "Not since Lewis Carroll wrote about logic has an expert produced a textbook on logic so saturated with delightful problems, paradoxes, jokes, and philosophical implications

as Professor Smullyan's *Logical Labyrinths*. Opening with a superb assortment of his famous logic puzzles, Ray moves sure-footed through higher-order logics to the paradise of Cantor's alephs and Gödel's undecidables. If you've ever wondered why in first-order logic the statement 'All unicorns have five legs' is assumed true, Smullyan will make it all clear. It's a volume only Ray could have written, and it will be used in college classrooms for a long time to come."

I hope you will all forgive my immodesty, but really now, can you honestly blame me for being proud of this? Incidentally, AK Peters has done a really excellent job with this book. It is beautifully formatted and printed, and has a very attractive cover. My deepest thanks to the publisher!

NOTE. I SENT THE ABOVE MESSAGE, AS WELL AS THE POSTSCRIPT BELOW TO MANY OF MY FRIENDS. HERE IS THE REST:

P.S. I should also mention that in the coming year I am coming out with four more books:

(1) *Rambles Through My Library.* This is a literary anthology with my commentaries and contains a good deal of discussion of Chinese philosophy, poetry, and paintings. Publisher: Praxis International, Inc.

(2) *Reflections: A Religious, Philosophical, and Transcendental Journey.* Also to be published by Praxis International.

(3) *King Arthur in Search of His Dog and Other Curious Puzzles.* To be published by Dover Publications. Dover is also reprinting two of my puzzle books, *The Lady or the Tiger?* And *Satan, Cantor and Infinity.*

(4) *In Their Own Words — Pianists of the Piano Society.* I have spoken about this book earlier in this work.

R.: I recently had the following exchange with my second cousin, the pianist Jacob Smullyan, which I believe would be of interest to you.

Jacob wrote me:

"I spent some time recently reading Schopenhauer, whom I discovered was a very different kind of writer than I had imagined him to be, very full of life and wit and insight, despite his pessimism. I had imagined someone dull, pompous, and morbid, and he is really none of these things. Even the pessimism somehow avoids morbidity, because the more vehemently he asserts that he is a pessimist, the less I believe him! Anyway, in re-reading *The Tao is Silent*, I found many ideas that reminded me strongly of Schopenhauer, as if, still under his influence, everything I read tended to be automatically translated into Schopenhauerian terms, almost like seeing an after-image after staring for some time at an object. How easily our thought may be distorted in that way. Nonetheless, there is some connection. (Inevitably, as he was so sympathetic with Eastern philosophy.) I can almost imagine him figuring in one of your dialogues, as he would agree with you on almost every point, but for largely temperamental reasons arrive at a totally different (and less desirable) place through it."

I wrote back:

"Hi, Jacob! I am also very fond of Schopenhauer. What you said is remarkably similar to a conversation I once had with the philosopher Bausma. I asked him, 'Why is it that when I read the pessimistic philosophers, instead of being depressed, I feel elated?' He replied, 'Of course, because you know it isn't true!'

I like to make a distinction between what I call ESSENTIAL pessimism and CONTINGENT pessimism. By the former I mean the doctrine that existence is necessarily painful, whereas by the latter, I mean the doctrine that existence is as a matter of fact painful, but doesn't necessarily have to be. For example, Schopenhauer was the former when he said that as soon as one satisfies a desire, another one arises. He was the latter when he complained about people of this world being so mendacious. It seems to me that if essential pessimism is true, contingent pessimism shouldn't matter — even if people were not mendacious, life would be horrible anyhow!

I believe that Schopenhauer's successor Eduard von Hartmann is far more interesting! I strongly recommend that you read his *Philosophy of the Unconscious*. It is really powerful stuff! He is truly an essential pessimist, and his idea of what we should do about it is utterly fantastic!

By the way, do you know the difference between an optimist and a pessimist? An optimist believes that this is the best of all possible worlds, and the pessimist agrees!

I once thought of the difference between an optimist and an incurable optimist: An optimist is one who says, 'Everything is for the best; mankind will survive.' An incurable optimist is one who says, 'Everything is for the best; mankind will survive, and even if it doesn't survive, it's still for the best!'

Best regards, Raymond"

R.: Here is a way to foretell the future. You wish to know whether or not a certain event will take place. You ask, "Will the event take place?" and you toss a penny with the understanding that heads means YES and tails means NO. Well, you toss and get a response, but how do you know if the penny is accurate? To find out, you toss a second penny and ask whether it and the first penny are alike, as far as their accuracy is concerned (that is, whether the two are both accurate or both inaccurate). Suppose it answers YES (coming up heads). Well, either the second penny is accurate or inaccurate. Suppose it is accurate. Then the two really are alike, as the second penny correctly testified, which means that the first penny is also accurate. Therefore, if the second penny is accurate, so is the first. Now suppose the second penny is inaccurate. Then, contrary to the false answer it gave, the two pennies are not alike, and since the second is inaccurate and the first is different from the second, the first penny is accurate! This proves that if the second penny answers YES, then, regardless of whether or not it is accurate, the first penny must be accurate.

A similar analysis, which I leave to you, reveals that if the second penny answers NO, then the first penny must be inaccurate. And so after the second toss, you will know whether or not the first penny is accurate, and so you will know whether or not the event will take place. Voila!

R.: In his helpful "Hints for Etiquette; Or, Dining Out Made Easy", Lewis Carroll (author of *Alice in Wonderland*) says two wise things:

(1) We do not recommend the practice of eating cheese with knife and fork in one hand and spoon and wine glass in the other.

Better still:

(2) As a general rule, do not kick the shins of the opposite gentleman under the table, if personally unacquainted with him: your pleasantry is liable to be misunderstood.

pianolady: Great — more funnies! (No. 2 cracks me up.) Keep 'em coming. ☺

R.: I am particularly fond of British humor. Here is a perfect example: An announcer for the British Broadcasting Company was about to introduce a modern British composer and gave a brief biographical sketch. With a perfectly straight face, he said, "He started out in life as a dentist, and then decided to see if he could extract a living from music."

Here is another lovely example.

A detective was testifying to a judge about a matter pertaining to chemistry. After he finished, the judge said, "After having heard what you said,

I am no wiser than before!" The detective replied, "That may be true, Your Honor, but you are better informed."

R.: A lovely Chinese proverb: When the wrong person does the right thing, it usually turns out wrong.

pianolady: How about "When the right person does the wrong thing, it usually turns out right?"

R.: Pianolady, I would say NO.

I love Chinese poems. Here is one of my favorites:

> In the evening years of my life, given over to quietude,
> No longer a slave to the world's affairs,
> My future? I have no better plan than to retreat to my old forest.
> There the pine wind will play about my girdle,
> And the moon will smile at me as I play the lute.
> You ask what laws govern failure and success!
> Just listen to the fisherman's song drifting up from the deep river
> estuary.

R.: I like Ambrose Bierce's definition of a piano: A parlor utensil for subduing the impenitent visitor. It is operated by depressing the keys of the machine and the spirits of the audience.

Do you remember Bierce's definition of an egotist? One who thinks more of himself than of me.

The following story is true: The great American architect Frank Lloyd Wright was once testifying about something in court. The judge said, "You are an architect; is that correct?" Wright replied, "No, what is correct is that I am the world's greatest architect." After the court adjourned, a friend of Wright asked him, "Why did you say that?" Wright replied, "I had to. I was under oath to tell the truth."

A man told his friend that he had not spoken to his wife for 18 days. When the friend asked why, the man replied, "I didn't want to interrupt."

I just read the following delightful story about Liszt: One of his lady students was playing something for him and in it were two runs, and after each run were two staccato chords. She did the runs most beautifully, and struck the chords immediately after. "No, no," said Liszt, "After you make a run you must wait a minute before you strike the chords, as if in admiration of your own performance. You must pause, as if to say, 'How nicely I did that.'"

pianolady: I love that. Liszt was great!

juufa72: You should record some of his works. If you can play Chopin and his ballades, you can play Liszt. I know that you have the talent to make grown men weep.

pianolady: Oh, juufa72, you are a sweetheart! 😊 Unless you mean I play so badly that I make grown men weep? 😕 😊

R.: I recently thought of the following joke:

> *A*: I am a specialist.
> *B*: You specialize in what?
> *A*: Everything.

Someone recently invented a truly wonderful copying machine. It does double duty: as it makes a copy, it shreds the original.

Don't delete this just because it looks weird. Believe it or not, you can read it.

> Cdnuolt blveiee taht I cluod aulaclty uesdnatnrd waht I was rdanieg. The phaonmneal pweor of the hmuan mnid, aoccdrnig to a rscheearch at Cmabrigde Uinervtisy, it dseno't mtaetr in waht oerdr the ltteres in a wrod are, the olny iproamtnt tihng is taht the frsit and lsat ltteer be in the rghit pclae.

> The rset can be a taotl mses and you can sitll raed it whotuit a pboerlm. Tihs is bcuseae the huamn mnid deos not raed

ervey lteter by istlef, but the wrod as a wlohe. Azanmig huh?
Yaeh and I awlyas tghuhot slpeling was ipmorantt ...

Teacher:	Johnny, if your father had ten dollars and you asked him for six, how many would he have left?
Johnny:	Ten.
Teacher:	You don't know your math!
Johnny:	You don't know my father!

Question:	What kind of mathematics do pigs study?
Answer:	They study swines and co-swines. It's all part of pigonometry.

A curious fact:
$$111,111,111 \times 111,111,111 = 12,345,678,987,654,321.$$

R.: A riddle: What musical form is very frightening?

pianolady: Anything with the word "Vivace" or "Presto" on the top. 😊

R.: I thought the most frightening musical form is the scare-tso.

Next riddle: What composer always interfered in other people's affairs?

pianolady: Don't give the answer yet, Raymond. Still thinking on this one.

R.: I'll give a little hint. Instead of saying that he would INTERFERE in other people's affairs, let me way that he would MEDDLE in other people's affairs.

pianolady: Ohhhh, Mendelssohn?

R.: Just about, but it should be Felix Meddlesome.

pianolady: haha — good one!

R.: Consider the following sentence:

I COULDN'T FAIL TO DISAGREE WITH YOU LESS!

Does this mean that I agree with you or disagree with you?

pianolady: It means that you "disagree". If you couldn't fail to disagree, that means you really are successful at disagreeing.

R.: Couldn't fail to disagree does indeed mean disagree, but couldn't fail to disagree LESS means couldn't disagree less, i.e. agree. Or at least that's how I see it.

Also, think of it this way. Since two negatives make a positive so does any even number of negatives and there are 4 negatives in the sentence, namely, "couldn't, fail, disagree, less".

I just heard a good joke: A fifty-year-old man had a wife the same age. A fairy appeared to him and said that she would grant him one wish. He said that he wanted a wife 30 years younger than himself. She said OK and waved her magic wand and he suddenly became 30 years older.

A rich but stingy man tried to get into Heaven. St. Peter asked him what he ever did for anyone. He replied that he once gave a nickel to charity and once gave a nickel to the Salvation Army and recently gave a nickel to a beggar. St. Peter asked God, "What should I do with this guy?" God replied, "Give him back his fifteen cents and tell him to go to Hell!"

Just heard a cute one: What composer is very fragile? Answer: Benjamin Brittle.

R.: Here is something I would like to share with you. At a party I met a certain concert pianist, whose name I will not mention. I asked him whether he would play something for us. He replied, "I play only at concerts. I don't like parlor pianists!"

By contrast, Alicia de Laroccha once visited my wife and me for the purpose of letting me play for her the program that I was soon to play at a concert at Rockefeller University. After hearing me play and making some helpful suggestions, she played for us the whole evening, and seemed extremely happy doing so! I will never forget how beautifully she played the Schumann Humoresque.

I was so terribly sad to hear that she passed away last month. She was such a wonderful person.

pianolady: Oh, Raymond, I didn't know you once played for Alicia de Laroccha and that she had played for you! Wow! I've been listening to her recordings practically every day. She was amazing.

P.S. I wish you would tell us the name of the other pianist — maybe you could tell me in private?

R.: I like the comedian who said to a friend, "I defended you the other day. Someone told me that you were not fit to live with a pig, and I said you were."

A man told his friend, "God doesn't answer prayers. I prayed for a million dollars and I never got it!" The friend replied, "God did answer your prayer. The answer was NO."

I came across the following interesting true incident. A depressed man went to a psychiatrist and was telling him how sad all of life is. The psychiatrist said, "I suggest you go and see Gibraldi the clown. He should cheer you up!" The man said, "I am Gibraldi."

R.: Definition of a shotgun wedding: A matter of wife or death.

I just came across the following true story. A man met a friend on the street and asked him if he knew the date. The friend replied that he didn't. The man said, "There's a newspaper sticking out of your pocket." The friend responded, "No good, it's yesterday's."

One person asked a friend whether it was better to marry or to remain single. The friend replied, "Whatever you decide to do, you will always regret it."

R.: Here are some cute puns I found on the Internet:

(1) The fattest knight at King Arthur's round table was Sir Cumference. He acquired his size from too much pi.

(2) I thought I saw an eye doctor on an Alaskan island, but it turned out to be an optical Aleutian.

(3) She was only a whiskey maker, but he loved her still.

(4) A rubber band pistol was confiscated from algebra class, because it was a weapon of math disruption.

(5) No matter how much you push the envelope, it will still be stationery.

(6) A dog gave birth to puppies near the road and was cited for littering.

(7) A grenade thrown into a kitchen in France would result in Linoleum Blownapart.

(8) Two silk worms had a race. They ended up in a tie.

(9) A hole has been found in the nudist camp wall. The police are looking into it.

(10) Time flies like an arrow. Fruit flies like a banana.

(11) Atheism is a non-prophet organization.

(12) Two hats were hanging on a hat rack in the hallway. One hat said to the other, "You stay here. I'll go on a head."

(13) I wondered why the baseball kept getting bigger. Then it hit me.

(14) A sign on the lawn at a drug rehab center said, "Keep off the Grass."

(15) The midget fortune teller who escaped from prison was a small medium at large.

(16) The soldier who survived mustard gas and pepper spray is now a seasoned veteran.

(17) A backward poet writes inverse.

(18) In a democracy it's your vote that counts. In feudalism it's your count that votes.

(19) When cannibals ate a missionary, they got a taste of religion.

(20) If you jumped off the bridge in Paris, you'd be in Seine.

(21) A vulture boards an airplane carrying two dead raccoons. The stewardess looks at him and says, "I'm sorry, sir, only one carrion allowed per passenger."

(22) Two fish swim into a concrete wall. One turns to the other and says, "Dam!"

(23) Two Eskimos sitting in a kayak were chilly, so they lit a fire in the craft. Unsurprisingly it sank, proving once again that you can't have your kayak and heat it too.

(24) Two hydrogen atoms meet. One says, "I've lost my electron." The other says, "Are you sure?" The first replies, "Yes, I'm positive."

(25) Did you hear about the Buddhist who refused Novocain during a root canal? His goal: Transcend dental medication.

(26) There was the person who sent 10 puns to friends, with the hope that at least one of the puns would make them laugh. No pun in ten did.

pianolady: Some of those are real "groaners" and some are real funny. I must be in a weird mood because I laughed out loud at No. 7.

R.: I just came across the following delightful incident about the conductor Eugene Ormandy: As a child, he was once taken to a violin recital. He was so disturbed by a wrong note that, from his seat, he yelled out, "F sharp, not F natural!"

I love the incident when Godowski was giving a piano concert and Vladimir de Pachmann was in the audience, and at one point, while Godowski was playing, de Pachmann went up on the stage and pushed Godowski aside, and said, "No, no! It should be played THIS way!"

juufa72: Time for a puzzle: Laura said to Tim, "Two days ago I was 7, but next year I'll be 10!" She was telling the truth! How is this possible and what day was it?

pianolady: Hmmmmm ... leap year?

R.: I also thought of Leap Year, but I don't see how that helps. I give up!

pianolady: Wait ... wait ... wait .... (Good time of the year for this one, juufa72.) The day must be January 1ˢᵗ. You were 7 years old two days ago on Dec. 30th. Then on December 31st you turned 8. Since it is now the start of a new year, you will be 9 on the same day this year that you turned 8 last year (a few days ago). And so the following year you will be 10.

R.: That's an excellent problem, juufa72. Where did you get it?

juufa72: Found it in a child's book about mathematics and fun with numbers. Though, I tweaked the problem to make it a little more difficult for our adult minds! (The original started with "On January 1st ... .")

R.: Here is a puzzle of mine you might like:

The Case of the Smithsonian Clocks

Two friends, whom we will call Arthur and Robert, were curators at the Smithsonian Museum. Both were born in the month of May, one in 1932 and the other a year later. Each was in charge of a beautiful antique clock. Both of the clocks worked pretty well, considering their ages, but one of them lost ten seconds an hour and the other gained ten seconds an hour. On one bright day in January, the two friends set both clocks right at exactly 12 noon. "You realize," said Arthur, "that the clocks will start drifting apart, and they won't be together again until ... let's see ... why, on the very day you will be 47 years old. Am I right?" Robert then made a short calculation. "That's right!" he said.

Who is older, Arthur or Robert?

Respondent: I'm still not sure about this, and I'm not even looking it up on the Internet, but I will answer it in a simple form: Arthur is older because his clock is going backwards by 10 seconds at a time, and Robert's clock is going ahead by 10 seconds at a time. So then Robert's clock will meet up with Arthur, who is already 47 and so at the time when the two clocks meet, it will be Robert's birthday, and he will turn 47.

R.: No, that's not right.

Solution to the Smithsonian Clock Puzzle

Each hour the two clocks drift 20 seconds apart. We must first determine in how many days they will come together again. Well, they will be together again after they have drifted 12 hours apart, which is 43,200 seconds (12 × 60 × 60). Since they drift apart 20 seconds each hour, they drift apart 43,200 seconds in 2160 hours (43,200 divided by 20), which is 90 days (2160 divided by 24). Thus they will be together again in 90 days. Now, how can one fit 90 days between some day in January and some day in May? Only from the last day in January to the first day in May, and provided IT IS NOT A LEAP YEAR. [You can check this on a calendar.] In a leap year it is impossible for any day in January to be exactly 90 days away from any day in May. Thus the year of the story is not a leap year! Now, since in the year of the story, Robert is 47, had he been born in 1933, the year would be 1980, which IS a leap year. Hence Robert was not born in 1933. Thus he was born in 1932 and is therefore the older one.

R.: "Never lend books, for no one ever returns them. The only books I have in my library are books that other folks have lent me." Anatole France quote.

Another quote of Anatole France that I like is, "If a million people say a foolish thing, it is still a foolish thing."

I once saw a cute cartoon of two rats in a maze, where one said to the other, "Boy, do we have those psychologists conditioned! Each time we run the maze they give us food!"
P.S. Does the name Pavlov ring a bell?

Just received a cute quote! Karl Barth's surmise: "While the angels may play only Bach in praising God, among themselves, they play Mozart."

Having recently read about the pianist Ruth Laredo, who was a good friend of my friend, the late Leon Kirchner, I could not resist coming up with the following:

RUTH LAREDO

The lovely pianist Ruth,
Whose playing revealed the truth,
Of composers both great
And composers both small.
She really did justice to all!

R.: Another musical riddle: What French composer is like liturgical music in the key of A?

pianolady: I know … it's Debussy, Ravel, Poulenc, Fauré. 😊

R.: Pianolady, I get the Fauré, but not the others there. I had in mind Mass-in-A (Massenet).

pianolady: Good one, Raymond!

R.: A sad story: Once upon a time there was a man so charming that all the ladies fell in love with him. His wife was extremely jealous. Her jealousy once reached such a pitch that she prayed to God that he would lose all his charm. Her prayer was answered. As a result none of the ladies ever loved him again … including his wife.

R.: A SHAKESPEAREAN MONICA

To be, or not to be,
In love with Lady Monica,
That is the question!
Whether 'tis nobler in the mind to suffer
The slings and arrows of unfulfilled love,
Or to take arms against a sea of troubles
And by opposing end them.

To die — to sleep,
To dream of Lady Monica!
'Tis a consummation devoutly to be wished.
But then to wake, and be with her no more!
Ay, there's the rub!
Tomorrow, and tomorrow and tomorrow.
Grows Monica more lovely from day to day!
To the last syllable of recorded time,
The way to the lady's heart.
Out, out, oh Chopin!
Whose music Monica so dearly loves!
Life's but a musical tribute,
That has its hour upon the stage,
And then is heard no more. 'Tis a tale
Told by a piano lady
Full of sound and beauty,
Signifying all that is wonderful!

10 | Lovely Ladies I Have Known

I have been an incorrigible flirt all my life. I happen to be a born-again romantic. I can't help it. It's genetic!

While my wife was still alive, someone once asked me, "Doesn't your wife object to your flirtatiousness?" I explained that my wife knows me too well for that. She knows that I am indeed a ladies man, but not a womanizer! She knows that my flirtations go no further than flirtations.

In an article I once wrote for a magic magazine, which I titled "Flirtatious Magic", I expressed my philosophy of flirting. I classified flirting as of two types — *seductive* and *complimentary*, and that I was interested only in the latter. My idea of a successful flirtation is when the lady says something like, "You are very kind," or "You are very sweet," or "You've made my day."

Now let me introduce you to some of my lovely ladies!

I have already told you about my lady editor Ann Close, who is surely one of the lovely ladies I have known. Another lady editor of whom I'm very fond is Rochelle Kronzek, an editor at World Scientific Publications, and a former editor at Dover Publications. Rochelle edited some of my books published by Dover. She has been most friendly and helpful, and is now shepherding this book through publication. She has visited me more than once, and has met my dear friends Sylvie and Wayne. The four of us had a great time together.

I once invited Monica Alianello, the "pianolady" of the previous chapter, and her family to visit me at my house in the beautiful upper Catskills. The sons, however, had different plans. Monica wrote me her regrets and added, "I'm sure you would be fun to be with!"

Another lady of the Piano Society who makes me wish I were younger is Tania Stavreva. In addition to her specialty, which is modern music, she plays classical music beautifully! I was very moved by her performance of Beethoven's Sonata Op. 109! Our relationship is nicely described in her own words:

Tania Stavreva

I met Raymond Smullyan in 2008 in Boston. I became a member of the Piano Society, where I shared my recordings. Raymond really wanted to stay in touch and hear me play live as well. He invited me a few months later to a piano gathering at a private house in Boston (back then, I was still living in Massachusetts). I had a lot of fun because all the people were very nice and very warm. Usually when I am surrounded by a lot of pianists (at competitions, festivals, auditions, etc.) the atmosphere is very competitive, but at this piano gathering I enjoyed meeting both professional and non-professional pianists, as well as music lovers from different ages and backgrounds. This was also the first time I heard Raymond play live. His playing (bearing in mind he is in his nineties) has no comparison to the energy, dynamism and virtuosic technique, of, let's say, a young prodigy, but his feeling, his passion and love for music is something I will never forget. The way Raymond played Bach for us at the gathering is the way I consider what music should be about — loving, sharing, and connecting to the audience.

I often listen to Raymond's videos when he posts on Facebook and YouTube, and I remember his Schubert and Scarlatti very well. When I listen to his performances, I can feel his heart beating. I cannot skip the video or do something else while listening, because being able to see and hear the honesty of his playing and how fully he expresses himself through music ... this is something I don't see always even at Carnegie Hall, and I appreciate immensely when a performer is true art.

I was also honored to visit Raymond at his home in Elka Park, NY (I think it was in the summer of 2010). He has a beautiful grand piano there, and I had a lot of fun playing on it. I am going to share a funny story from that visit. I might have been the first pianist to inspire him about modern music (20th and 21st century). He came to my Carnegie Hall debut in 2009, where I performed works by Ginastera, Shohat, Scriabin, Debussy, Vine and A. Vladigerov. Raymond is a big fan of more traditional composers such as Bach, Beethoven, Chopin, Schubert and Schumann, so at the first piano gathering (the one in 2008) I played Beethoven Op. 109, some Chopin études and Ginastera. Raymond didn't like the Ginastera the first time so much, but after hearing it a few more times, and then after my Carnegie Hall debut, he said that he was really enjoying the 1st Piano Sonata, Op. 22. Another modern work he liked was Alexander Vladigerov's Variations on a Bulgarian Folk Song, "Dilman, Dilbero". So, now getting back to the funny story — when I was visiting him at his house and we talked about modern music and newer styles, I decided to show him some rock music! I personally enjoy many musical styles: classical music (of course), but also listen to jazz, blues, rock and even some metal. I decided to introduce him to the rock band Queen. We found a YouTube video, and he loved when Freddie Mercury was playing the Bohemian Rhapsody on the piano. But when the electric guitar came … NO! We had to stop the video; he just couldn't handle the distortion sound of the guitar.

To conclude, I'd like to say that Raymond is one of the most fascinating people I've met on planet Earth! He is noble, pure, intelligent, a genius mind, and I am thrilled that I had the opportunity to share music with him, be friends with him, learn some logic quizzes from him, meet him and even introduce him to some newer modern composers of the 20th and 21st centuries. Raymond also supported me as an artist during a difficult period of my life, and he helped me move forward and continue to create and develop in my career. I am grateful, and words are not enough to describe how much I thank him for the help and support! I wish him many, many more years of happiness, musical gatherings, good health and continuous success!

I have already told you how photographs of three ladies of the Piano Society made me wish I were much younger. As a matter of fact, *all* the ladies of the Piano Society have that effect on me! Now I will most joyfully tell you about some other lovely ladies I have known, who are not members of the Piano Society.

Irene Schreier Scott

Irene is a pianist, but not a member of the Piano Society. She is the daughter of the famous mathematician Otto Schreier, and is married to the eminent logician Dana Scott. I have known her since she was 10 years old. She is a lovely lady, indeed!

> The illustrious Dana Scott
> Thought much of theology was rot.
> But he never did tire
> Of Irene Schreier,
> And so they tied the wedding knot.

> As for Irene Schreier,
> Whom all musicians admire,
> And who later became Mrs. Scott,
> She plays the piano a lot,
> All music from Bach to de Falla,
> And her playing has plenty of fire!

Kate Jones

> The Puzzle Lady Kate Jones
> Invented many fine koans.
> Ah, yes! This remarkable Kate,
> Simply could not wait
> To positively rate
> Her good friend Ray,
> Who in turn would say,
> "Her puzzles are really quite great!"

Kate Jones is a designer of games and puzzles and is the head of Kadon Enterprises, which can be found on the Internet. I first met her at a gathering honoring Martin Gardner. At one point I told her that I would bet her that I could kiss her without touching her. She immediately said, "I know; you will lose!" When I congratulated her for her cleverness, she told me that she really did not deserve all the credit, since she had recently heard a similar one — the one about the man at a bar who bet his friend one quarter that he could drink his friend's cocktail without touching it.

A lovely friendship between the two of us developed over the years. In Kate's own words:

> For love of Ray, a ray of love
> (recital of a sonnet that does not require a title)
>
> When legends walk among us, gleaming bright,
> And talk in paradoxes steeped in light,
> With rapier logic looped in nooses tight,
> A checkerboard of interwoven sight,
> When, as I say, such beings grace our plane
> And test the limits of what makes us sane,
> Their stunning talent triggers mankind's gain —
> And o'er them all zooms Raymond Smullyan's brain.
> From myth to math to magic, music, mind,
> His equal in these matters you will not find.
> His fingers stroke the keys like lovers 'twined.
> While twinkling eyes seduce, serene and kind.
> What man is this, in all of space and time,
> Who thinks a thought that doesn't think, in rhyme?

I met Raymond Smullyan several times in the Gatherings of Gardner in Atlanta, where he also played piano in the lobby with those incredibly beautiful hands of his. I knew he was a legend even before that, from his books and what Martin Gardner had told me. So it was delightful to find him a flirtatious old codger who declared more than once that he would have married me if I were not already married. On occasion he would telephone me to ask advice, usually involving a romantic interest. We emailed charmingly and teasingly — it breaks my heart that a computer failure has wiped all those files — and wrote poems to each other, with Ray's witty paradoxes and logical mind traps. He is a total dear, and I want him to live forever! He was kind enough to mail me a CD honoring his wife, who lived to be 100. And, of course, I have many of his books in my library.

What more can I say about a living legend, hugging whom has been one the most memorable pleasures of my life? There is a photo of me with Ray on our website (www.gamespuzzles.com/raymond.htm), my zero degrees of separation from one of the greatest minds of two centuries.

My little sonnet for him alludes to the many fields in which he has left his incomparable footprints. Only Ray will be able to read between the lines to spot my wicked but affectionate humor.

Here is a poem I sent her and her response:

A SAD SONG

Alas, oh lovely Kate,
I'm afraid it is my fate
To never have a date
With any girl as great
As the incomparable Kate.

'Tis a thing I really hate,
That, however long I wait,
It simply is too late
For me to have a mate.

From Kate:

Dear Raymond, yes, you sent it,
And then revised and bent it.
Your longing would resent it
And your humor would relent it,
So may you not repent it!

It's not the dating that you miss
Nor the longing for a kiss.

It's the mating game, of course,
Whose absence brings on your remorse.
But you're not really all alone:
A million friends and fans are shown.

— Kate, the great

Teja Krasek

In her own words:

It was in Atlanta at the seventh Gathering for Gardner, the conference initiated by Tom Rogers in honor of Martin Gardner, the great popularizer of mathematics, when I noticed this tall man with a white beard and white hair walking energetically through the lecture hall as if in a hurry. I'm an artist who is interested in mathematics, and I asked my mathematician friend sitting next to me who the man was. I was told, "This is Raymond Smullyan." Wow! I had previously heard of Raymond, the mathematician, pianist, logician, philosopher and magician — yet I never expected to see him in person.

As you might know, the Gathering of Gardner conferences are tightly packed with interesting talks by genius mathematicians and puzzlers, with many social activities, not to mention breath-taking magic and other performances. Amidst all this chaos, I did not get a chance to talk with Raymond.

So I found myself enriched with tons of new knowledge and a big bag of awesome G4G7 exchange gifts at the Atlanta airport heading back to Europe.

However, the universe is sometimes rich with pleasant surprises. One of them was that Raymond and I appeared next to each other in a seemingly endless security check queue, and we started a conversation. I told him we had attended the same conference, and we talked about our flight destinations and so forth.

By then, for some years very tight security measures had been carried out at airports. It made me really sad to see how Raymond, this adorable man, who really, really did not represent any trace of threat to anyone, had to take the belt off his trousers, take off his shoes, carry them while walking, etc.

While Raymond did not bother to complain, my heart ached to see how this noble man in his 90s, whom I deeply admire, had to go through the hassle of security check procedures. If anyone would ask me, it would be Raymond, with his precious, kind personality, his brilliant mind, and romantic soul, who would have protection in the form of some strong muscle-rich body guards, so that nothing would happen to him and no one would bother him!

I assisted him a bit, and eventually we were happily through everything. As we expected, none of us was seen as any kind of threat.

The nature of things, in these time-consuming security check procedures, a shared intimacy of walking together in our socks only, and perhaps a bit of the nature of our personalities, contributed to the fact that by the time we found ourselves inside the airport shuttle Raymond was generously giving me the most amazing poetic compliments. Every woman on Earth should be exposed to such poetry at least once in her lifetime!

Two years later there was another Gathering of Gardner conference, and once more I was privileged enough to listen to Raymond's incredible piano playing and to talk with him some more. Given that I'm an artist, a painter, if we're precise, I'm excused from having an Erdös number[5] of any kind. However, I'm immensely happy to have a Smullyan number of one! This doesn't mean that I coauthored a mathematical paper or proved some mind-boggling mathematical theorem together with Raymond. No, it simply means that Raymond in his eternal entertaining and playful spirit successfully performed on me one of his famous kissing logical puzzles, where he "lost" a bet to me and therefore, as a final result — kissed me on my cheek. This makes me a privileged and happy Smullyan number one, who travels in this world carrying Raymond's famous magical kiss, and I am not the only one.

Thank you, my dear Raymond, for being such a unique diamond, for enlightening our minds and hearts with your endless wisdom, humor, magic and kindness.

From Raymond:

Ay, yes, Teja the fair.
Spread the good word,
All of us have heard
How wonderful she is!
All the men love her,
None are above her,
Everyone flocks to her lair.

---

[5]This is a reference to Paul Erdös, a prolific mathematician. He wrote so many papers with so many coauthors, that having a low Erdös number became an informal measure of status among mathematicians. An Erdös number of one means you coauthored a paper with Erdös, a number of two means you coauthored a paper with someone who had coauthored a paper with Erdös, etc.

From Teja:

> Raymond, dear! You're incredible, thank you so much!
> You're the brightest shining Ray du Monde!
> And you certainly know how to make the members of the opposite sex
> blush! 😊😊

Cliff Pickover, a famous maker of puzzles, says (on The Official Teja Krasek Appreciation Page on the Internet):

> Teja Krasek is my favorite living European artist. One of her goals is to unite science, mathematics and art. Her works have been featured at international exhibitions. My colleagues consider her Eastern Europe's "MC Escher" and Slovenia's "gift" to the world. Escher thought big. Buckminster Fuller thought big. But Krasek outdoes them both. I've known her for years, and we both agree with Sven Carlson's statement: Art and science will eventually be seen to be as closely connected as arms to the body. Both are vital elements of order and its discovery. The word "art" derives from the Indo-European base "ar", meaning to join or fit together. In this sense, science, in the attempt to learn how and why things fit, becomes art. And when art is seen as the ability to do, make or portray in a way that withstands the test of time, its connection with science becomes more clear. Teja's favorite equation is
>
> $$\varphi = \frac{1+\sqrt{5}}{2} = 1.618033989.$$

From Raymond:

> Beloved Teja,
>
> I have been in correspondence with Cliff Pickover. Of course I talked a good deal about you and described you as "infinitely lovely". To my great delight, as part of his reply, he said, "Yes, we seem to have many things in common," which concluded with "AND an infinite fondness for Teja."
>
> How happy I am that he also so appreciates you!
>
> Love,
> Raymond

From Raymond:

Good Heavens! After seeing your photograph, I must ask you, "Are there any men in Slovenia who are NOT in love with you?"

From Teja:

Ooh, my dear Raymond! Thank you for your extremely kind words! 😊
Love your sense of humor!
Ha, would not know the answer to your profound question but I think
We should not exaggerate. 😊

From Raymond:

Dear Teja, I cannot tell you how excited I was when I read in your message that you were once thinking of being a concert pianist! Do you still play? Are you aware that I would have been a concert pianist had I not developed tendonitis in my right arm? I still play, and I make musical videos and put them on YouTube. Have you seen (heard) any? If not, the link is: www.youtube. com/profile?user=rsmullyan#g

P.S. It just occurred to me that if I should ever see you a second time, it would be a case of Teja Vu! [Have you heard that pun before?]

From Teja:

Dear Raymond,

I do still practice piano a little bit, as much as hands permit; my right hand was, as you know, injured and in October I was still not able to do anything, not even writing with it. Yes, I know, dear Raymond, that you were supposed to become a concert pianist, and sadly tendonitis prevented you from becoming a professional one! I'm happy you still play, and yes, I immensely enjoy the musical videos you have on YouTube! But it's in a way good we did not become concert pianists — we would probably not be invited to attend Gathering for Gardner in this case. 😊

From Raymond:

> Indeed, if I were 50 years younger at the PRESENT time, I would immediately fly over to Yugoslavia, put you in my suit case, take you to America, and marry you on the spot!
>
> All my love,
> Raymond

From Teja:

> My dearest Raymond,
>
> Oh, dear Raymond, I love what you've said, and I'd love that.
>
> You must know that your previous kind letter moved my heart to tears. And I did understand and felt that you meant PRESENT when you said you wished to be 50 years younger. Cannot tell you how much you moved my heart. You are such an exceptional and precious human being, and I feel so privileged.
>
> With love,
> Teja

Malgosia Askanas

Malgosia Askanas, a very intelligent girl from Poland, wrote a brilliant Ph.D. thesis under me about truth and provability. She also took my graduate course in set theory. At the end of the course, I gave oral final exams to my students separately. The exam I gave to Malgosia was unusual, to say the least! I asked her what her favorite theorem of the course was. She told me. I then asked her if she knew the proof. She said she did, and so I gave her an A! The whole point is that I knew her well enough to know that she perfectly knew what she knows and what she doesn't. Some students have no realistic notion of what they know and what they don't. But I knew that if Malgosia believed she knew something, then she really did!

Many years later, when I was teaching at Indiana University, Malgosia and her boyfriend visited my wife and me for several days. I had a lot of fun with my students in those days. Malgosia sat in on the last day of my graduate course in set theory, which Malgosia had taken with me several years earlier. I said to the class, "Dr. Askanas had this same course with me some time ago. However, there was one theorem in set theory that was not included in her course, which I will state and prove today. If you all feel that Dr. Askanas understands the proof, then I will allow her to keep her Ph.D.; otherwise I will rescind it! I will let you vote on the matter."

I then gave the theorem and proof. It was obvious from the discussion we then had that Malgosia understood the proof perfectly, and so the students unanimously voted that Malgosia should keep her degree.

Sue Toledo

Next, my former Ph.D. student, Dr. Sue Walker Toledo. She also wrote a brilliant thesis with me. She has subsequently done an excellent editing job on my manuscripts for *A Gödelian Puzzle Book,* for *A Beginner's Guide to Mathematical Logic* and now for this book as well.

We have recently had the following correspondence:

Dearest Raymond,

I'm mailing you today the next batch of chapters with my questions to you on them. But I want to write to you about something else first. It's been on and off my mind for a while, and especially so this week. I woke up this morning thinking about it.

It's about the *memoir* you told me that you are now writing or planning to write. As I work through this logic book and not only revisit old things I felt were very beautiful when I first met them but also find myself encountering a number of new things equally beautiful (the end of Chapter 9 being my most recent pleasure), I constantly feel grateful for having had the great fortune to have known you and to have studied logic and set theory with you. But also for being able to get to know many other sides of you — your music, your magic, your philosophical thoughts, your puzzle and chess books. Your magnificent ability to entertain people and give them joy in life. Your very personality!

I don't know what you plan to put in this memoir, because many of your books have been like mini-memoirs, in the sense of being an open exploration of a part of your life and thoughts.

I find myself really hoping you are considering writing a true (and as full as possible) autobiography. You are one of the few great personalities and contributors of the 20th century. I would use the word genius, but that makes most people uncomfortable when you ascribe it to them. Not that every page I read, and everything else I know about you doesn't show me that that word truly applies to you. And *you* have applied your genius in your life in such a way that it has only given others great pleasure, not made them feel inadequate!

I would love to see a real autobiography, a deep picture of how the person you have become slowly developed from that little boy born to your parents — all his joys and disappointments, the things that made the deepest impression, the things that brought him to change direction or add a new one. How certain things were able to engross his mind and give him the deepest of intellectual pleasure. This is not just something I would like to have personally, to come to know one of my closest friends better. I think it could be very valuable to the world in general, especially at this very difficult time for almost everyone around the world (difficult economically, politically, emotionally, spiritually … whatever the last word means for each individual). Although there is probably some genetic component to genius, I feel that the environmental component dominates, that is, that even normal children, if their curiosity is constantly aroused by the environment they find themselves in … and, above all, if they are given the time and material aid to follow their curiosity and even encouraged to do so … they too are likely to become geniuses. After all, most geniuses are not the children of geniuses, nor are most children of geniuses themselves geniuses. So I look forward to learning more about the environment you lived in, from birth until I was invited for the first time to your home and began to see personally the environment that was still sustaining you as an adult.

For I feel that your life and personality have been very beautiful ones. I feel that this world now is becoming less and less aware that people like you can even exist, that lives like the one you have had can ever occur. I sense that even the richest class of people, those who do have the freedom to live any way they wish, don't understand that your kind of path is a possibility. Very importantly,

I believe, people need to hear what *deep* intellectual pleasure is like. Very few get to have the opportunity to ever experience it … ever (a true tragedy, I believe, on the par with never having a deeply satisfying love relation).

Well, that was it. I pray that writing the kind of autobiography I long for would be something that would give you pleasure.

Now, what I'm mailing you today is all about small things, simple editing questions. The thing is, here I have a book that is written by a person who I feel is one of the best writers of all time, and I am giving him many editing proposals with the goal of improving little parts of his work. Worse, I am aware that sometimes the actual rewriting I propose, although it may solve the problem of something overlooked, is usually seen by me as not actually written as well as the surrounding text. So, *please*, even if you see the value of what I want to add, do rewrite it if you can improve on my suggestion! I also apologize for the tedious nature of the work I'm asking you to do (in checking all my proposals and answering all my questions). But the book is worth it, is the most beautiful logic book I've ever seen.

Thanks again for this wonderful work.

All the best, Sue

From Raymond:

Dear Sue,

Many thanks for your kind words. It is funny that you used the word "genius". As a matter of fact, in my youth I fantasized myself as one day becoming a "genius". Later I realized that my fantasy was unrealistic. Now, it would be false modesty on my part not to concede that I am multi-talented. Of course I am! But that is a far cry from *genius*. To me, a genius is one who had made major contributions to the world. I have never done that, though I know I have made numerous minor contributions. Several people have described me as a Renaissance man. This, though perhaps over-flattering, may well contain an element of truth.

As to the book I am writing, it is not so much an autobiography as a collection of some events in my life, combined with thoughts, jokes, anecdotes and some logic puzzles. I have already written a kind of autobiography titled *Some Interesting Memories*. Do you have a copy? If not, I'll see that you get one. Incidentally, this is the only one of my published books, that got a negative review, one which I found quite funny. I attach a copy.

My very best, Raymond

The attached REVIEW: Many years ago I got fascinated with Smullyan's *What is the Name of This Book?* and since, have been buying most of Smullyan's books. It turned out, however, that all those books have been strikingly similar (with the only exception of *5000 BC*). All the same logical paradoxes, jokes, and anecdotic accounts from Smullyan's life, which you do not perceive as funny after you read them again and again. This book beats all the previous Smullyan records on repeating the old jokes and puzzles from his own earlier books, or even older folklore jokes. The only new "original" content here is Smullyan's unending boasting of how talented he is as a musician and magician, and how highly other people think of him. I find this narcissism particularly distasteful (although the author himself modestly calls it "honest"). It would be much better for humanity if, instead of publishing this tasteless compendium, they reissued his marvelous *What is the Name of This Book?* and *5000 BC and Other Philosophical Fantasies*, which are long out of print.

From Sue:

Dear Ray,

I could say that you've given me the only evidence I ever had that you're not a genius, i.e. that you don't recognize it.

But that would be silly. There is of course a hierarchy of abilities, as well as of the accomplishments that people have managed to make in their lives, given their original natural abilities and the circumstances in which they lived their lives. Where to apply the label "genius" is impossible to specify. Our mutual friend Stanley mentioned to me a couple times that all of the young people with whom he had formed close bonds were geniuses, and that I was a genius. But since I had known so many of the most gifted mathematicians (and a few scientists) of the last century, and observed how very much more they knew about their fields than I did about mine, I of course dismissed this categorization.

Are you aware that studies have shown that anyone who masters playing difficult classical music on any instrument of any complexity (drums and chimes excepted perhaps) turns out to have developed a much larger area in their brain related to music than others? I would guess that would be true for anyone who has mastered a large amount of complex mathematics or physics as well.

But humans are very competitive (above all, the males of our species, but we females also have a bit of that demon in us). It is unfortunate for all of us that only a few can discover the most difficult and valuable new theorems (or whatever), because once a person finds that someone else has made his/her discovery first, their own rediscovery is no longer a big deal (to them *or others*), even if they did make their own discovery independently. My classmates and I learned topology at the University of Michigan by proving on our own theorem after theorem that had gained someone fame when they were the first to prove the theorem. When I had the opportunity to spend three years doing nothing but mathematics 12–16 hours a day, I discovered on my own a good number of results in number theory and one major one in combinatorics (as applied to statistics). I learned later that my results had all been published recently in major journals by other mathematicians.

In any case, the bar at which I start calling people geniuses is high enough that it doesn't allow many people in. But it's an amazing crew.

Let's just say that what I meant in terms of your accomplishments is that, for one thing, I am sure that it is unlikely that many over the course of history have been able to have so many high level accomplishments in so many different domains as you. Being a gifted musician, very high level mathematician, and an unbelievable entertainer all at once!!!! Your personality, too, is quite unique, and very beautiful. Don't you get it? You're *special*.

And you also have an extraordinarily valuable and *very rare* talent regarding which I doubt if anyone has been as productive as you have. You are able to take results already known, and find a much clearer way for all of us to understand them. And to see them from all sorts of new angles. You have lived in the *aesthetics* of mathematics (and of other fields as well ...), which I am sure has been a great pleasure for you, but which has also given great pleasure to many others as well.

And just forget about the genius stuff. You've had a great life that I believe people would enjoy sharing with you.

No, your book *Some Interesting Memories* is not one of yours that I have. I would of course love to read it.

With great affection, Sue

P.S. Just read the bad review. Wow! I have learned to my horror by looking at reviews of other writers I admire that many reviewers seem to delight in tearing down writers and their work, even while simultaneously expressing some admiration. Of course, since you have written so much during your life, and much of it is written as if you're taking the reader along with you through thoughts you've been having, there will be repetitions in almost anything autobiographical. But for one thing, I was thinking of an autobiography by you as potentially having many more readers than those who have been buying your books for years. And if you apologize to those who may have read some of the material in the past, you might find that most of those who've read you with pleasure will even enjoy revisiting some old jokes and puzzles. (I know I do.) As long as you also find some new things in your life to explore with them.

A Thanksgiving letter from Sue:

November 27, 2014

Dearest Raymond,

I want to send you a Thanksgiving letter today to give you thanks. For one thing, to tell you once more how grateful I am for you being in my life. You have helped me for decades, starting during my student days, when your courses gave me great pleasure (and a degree in the end), and when you also got the opportunity to help me out in a major crisis. This was when my ex-boyfriend was threatening my life; do you remember that incident?

But then now again, recently, with these wonderful books of yours to edit (which process helped wake up my mind, which had been getting lazy), and with the great financial help.

The other day I attended a celebration of the life's work of an interesting woman. Person after person came forward to tell her how she had literally saved their lives, to tell her how grateful to her they were for having made everything then good in their lives possible. As I thought of how fortunate this good, strong, wise woman had been to have found herself (by chance!) in a position to help so many people, I thought of you, and realized that you too have done that, even when you weren't directly intervening in a life.

I was trying to say these things in the letters I wrote to you earlier, but had focused more on telling you how grateful I was to you for your role in my own life, rather than on reminding you about the valuable role you have played in the lives of almost everyone who's come in contact with you or your work.

I'm talking about *your legacy*, about the writings and recordings that have already helped many in their lives and will continue to do so for new people in the future.

Did I ever tell you how just a single little math puzzle of yours managed to turn some children on to mathematics in one of the worst middle schools I ever subbed in? I told three boys I would tell them something surprising. They thought they had complete control over their friendships, but I could show them something that they couldn't control, as strange as that might seem. I said that if all the kids in the classroom at that moment got together to try to make it so that no two of them would have the same number of mutual friends, they wouldn't be able to do it, even if everyone was willing to give up some friends or make new friends in order to make it happen. (Problem #2 in your book *The Riddle of Scheherazade*). We talked long enough so that the children really did see why that was true. A couple hours later, a child I had never seen came up to me in the hall and said excitedly "You're the sub who can show a really surprising thing about friendship, aren't you?" Obviously the first three kids had really been turned on. By you! And perhaps they had been able to completely pass that pleasure on to a few others.

In fact, you've been able to touch more people than the woman who so many were celebrating here last week. You've given many people *deep* pleasure with the beauty of your mathematics. You've given people fun and have helped them develop their minds with your puzzles. You've made innumerable people happy with your (always kind-hearted!) jokes and with your magic. You've touched people deeply with your music. You've helped many realize that they can understand complex things. *You teach people that using their minds can be a great pleasure!!!*

I am extremely happy that you decided to write these recent books to round out your contributions, to try to get as much as you possibly can give to others out there in the bookstores and libraries and on Internet.

Many, many thanks, sweet man!!!! Happy Thanksgiving!

Sue

"Caroline"

My relation with Caroline started out so beautifully, and ended up so painfully! It is best that I do not use her real name, but shall call her "Caroline".

I first met Caroline many, many years ago when Blanche was still alive. Blanche and I had tickets to a recital in Carnegie Hall, but for some reason Blanche couldn't make it, and I was able to return her ticket, which was bought by Caroline. Thus, next to where I was seated was this tall, lovely looking girl in what otherwise would have been Blanche's seat. In Caroline's own words: "I sat next to this very distinguished looking gentleman, and longed to talk to him, but I did not want to seem too forward." It was I who broke the silence, and told her that she had purchased the ticket of my wife. She told me that she was a piano student from California and was looking for a teacher in New York. I then told her about my wife and her music school. She asked me whether my wife would teach her. I replied that she might, or more likely would recommend some other teacher, since she was acquainted with so many musicians in New York.

At the end of the concert, I took Caroline home in a cab and invited her to come to dinner and meet my wife, which she did. After dinner, she played a Beethoven sonata for Blanche and me, and Blanche recommended a teacher from a well-known music institution. Caroline went to the teacher, and the two got along beautifully. After some months of study, the teacher recommended she go to a well-known music school, which she did.

About five years later, I got a phone call from her. She asked me if I remembered her. At first I didn't, but when she told me of our meeting in

Carnegie Hall, I said, "Oh, of course I remember you!" I then told her which Beethoven sonata she had played for my wife and me in our home.

For about a year after that, we had numerous email and telephone conversations. In one telephone conversation I said something that I thought would be helpful, but it turned out to be the very opposite! I expressed a radical idea that shocked her so much that she screamed, hung up the phone, and wrote me that she never wanted to hear from me again, and that on no account should I ever contact her by any means whatever! She refused to read anything I wrote to her, telling a friend of mine that she dare not read anything from me, because, as she said, "Raymond is very clever, and might persuade me to trust him again, which would be very dangerous!"

Never in my life had I ever been so grossly misjudged! What she failed to realize is that even if my views are wrong, which is certainly possible, that does not mean that my motives were bad! I was trying to relieve suffering, not create it, as I believe she thought! If I was going about it in the wrong way, that was due to ignorance and misjudgment, not malice. Why couldn't she see that?

The situation was surely as painful to her as to me, because before our break, she seemed to think the world of me. So apparently I was now a fallen hero in her eyes!

My only hope is that one day — perhaps after my death, she will read this and she will then realize the truth of the matter.

Elizabeth

On a much happier note, let me now tell you about Elizabeth, a college senior I met one evening in a local restaurant where she was working as a waitress. She is extremely bright, and struck me as making an ideal lawyer! As soon as I saw her I asked her, "Why are you looking so extremely beautiful tonight?" She replied, "Because you're here!" After having flirted with her all during dinner, before I left I asked her, "Have you ever met any other man as dangerous as me?" She replied, "Yes, but not as suave."

The next day I came into the restaurant with a friend for lunch. I said, "Today I am even more dangerous than yesterday!" She replied, "That's unimaginable!" Before I left, I looked at her and sighed and said, "Oh, if only I were sixty years younger!" She replied, "If only I were sixty years older!"

Doesn't she sound like she would make an excellent lawyer? Well, a couple weeks later, in another restaurant, a very lovely looking lady, probably somewhere in her forties, approached me and said, "I understand you are infatuated with my daughter!" I said, "Good Heavens, are you Elizabeth's mother?" When she told me she was, I embraced her and gave her a kiss. She later told a mutual acquaintance that I kissed her in lieu of wanting to kiss her daughter, which was not at all true, since I found the mother extremely attractive in her own right. When I next saw Elizabeth, I told her how much I liked her mother, and added, "I thought I was in love with you, but since I met your lovely mother, I'm not so sure." She smiled, and later told this to her mother, who was equally amused.

The next time I saw Elizabeth was at the end of the summer, when she had to go back to school soon, and wouldn't be back until Christmas. I said, "Then the next few months I'll have to live with Elizabeth deprivation?" She replied, "I think you'll survive!" I was not sure I heard her correctly, and said, "Did you say that I would survive, or that you would survive?" She replied, "I said that you would survive." As she walked away, she added, "I don't know whether I will survive!"

Elizabeth is really quite impish! The last time I saw her was in the restaurant. She came over and sat opposite me, and with a twinkle in her eye said, "When are you and I going to elope?"

Betul Yilsnaz

My relationship with Betul was particularly beautiful!

Betul Yilsnaz is a graduate student in archaeology in Turkey. She once emailed me to ask whether my book *Satan, Cantor and the Infinite* was still available, and if so, where she could get a copy. I then mailed her a copy, and a great deal of email correspondence then ensued. I really have tons of back and forth emails from her, as well as from Teja Krasek, which I hope to someday publish elsewhere, but for now I wish to focus on one particular incident about Betul.

At one point in our correspondence, she wrote me, "I wish I would have the privilege of meeting you in person and listen to the secrets of having a long life love — like the one you are having with your dear spouse Blanche." I wrote back to her and told her that she used the word "having", which is the present tense. I told her that she evidently didn't know that my wife passed away several years ago. Then she wrote me something sublime:

Seemingly, concerning the word 'having' I need to elaborate why I used the word in present tense. To be precise, using the word in present tense was a deliberate choice simply because I do not see myself in the position of using the word in the past tense. I do not think a third party like me can be judgmental about YOU 'have' or 'had' that Love for your beloved spouse. Of course I know that she passed away, though does that mean that the Love that you 'have' or 'had' passed away with her? I always believe in aeon of love hence if the owner of loving heart is still beating for his or her love then loss of the other is just a continuum. I saw dear Blanche's picture in your biography and telling the truth, I got amazed by her radiating look.

* * *

Coming now to my present life, there are some other ladies (and men) that I am now in contact with.

First, fairly close to where I live are my dear Greek friends, George Billias, his lovely wife Matina and his daughter Athena. This lovely Greek family are my oldest friends of my neighborhood. Blanche and I spent countless hours with them. They play a major role in a video I made which is a documentary of Blanche's life.

I am constantly threatening to kidnap Athena. On one occasion I said, "Today I'm going to kidnap you!" She replied, "You've said that many times before. Somehow, you haven't been very successful!"

George Billias is very good at fixing things. When anything went wrong with my house, he would come over and fix it. I also keep telling George that I will run away with his wife! On one occasion when I threatened this, he said, "I know, and if anything goes wrong with her, you'll bring her back to me to fix!"

About a year or so before Blanche passed away, we met the lovely concert pianist, Lisa Kovalik, who is a professor of music at the Juilliard School. She is now a very close friend of mine. I also keep threatening to kidnap her, and when she asks me where I'll take her, I tell her, as I often tell others who I threaten to kidnap when they ask the same question, "To my secret love island in the South Pacific!"

My fourth contemporary lovely lady is Sylvie Degiez, who is married to Wayne Lopes. I met them sometime after Blanche passed away, and those two, together with Lisa, are now among my closest friends. Sylvie is a pianist and composer. Wayne is an extremely talented guitarist, an absolute virtuoso! I think of Sylvie and Wayne as my guardian angels. They take care of me in all ways imaginable. It took me a long time to figure out how Wayne got his name, but it finally dawned on me that it was because his love for Sylvie never wanes! [I love making puns. My doctor formerly had a lovely receptionist named *Amy*. I told her that she had that name because she was so amiable! She hadn't heard that pun before.]

Sylvie and Wayne and Lisa all live in New York City and have country homes up here near me in the Catskill Mountains. I see them virtually every weekend. Once I was at dinner at Wayne and Sylvie's house, and at one point, Wayne had to go out on some errand. I said to Wayne, "Do you dare to trust me to be alone with Sylvie?" He replied, "No!", and then added, "But I trust Sylvie."

Lisa, Sylvie, my second cousin Jacob Smullyan, and I are planning a joint concert I will soon tell you about, but first let me tell you about Jacob.

Jacob is a superb pianist. He has an extremely lovely wife and three wonderful, adorable children. This is one of the happiest families I know. I once put Jacob on the telephone with a friend of mine, and before Jacob took the phone, I said to my friend, "Jacob knows I'm in love with his wife, yet the two of us are still on speaking terms. Isn't that amazing?" Jacob then took the phone and said something that I thought was sublime! He said, "I'm glad that Raymond is in love with my wife. It shows he has good taste!"

Our planned concert has an interesting background. In 1929 a group of famous pianists got together to raise money to help another famous pianist who was in financial need. They put on a gala event at Carnegie Hall. The final piece of the concert consisted of all the pianists jointly playing the Schumann *Carnaval* in the following manner: The *Carnaval* consists of a series of short pieces. There were two pianos on the stage, and two pianists were initially seated. One of them played the opening piece, and when he was finished, the other one played the second piece as the first pianist walked away and a third pianist took his place. When the second pianist finished his piece, he was replaced by a fourth pianist, as the third pianist played his part. And so they alternated in this manner till the end. The concert was a great success! Well, it occurred to Lisa, Sylvie, Jacob and I to do the same thing, only instead of the Schumann *Carnaval,* we decided to do the Bach *Goldberg Variations.* And so we are practicing our parts. The only trouble is that my tendonitis has returned, making practicing quite difficult for me.

* * *

As I said at the beginning of this book, I now live alone in the beautiful upper Catskills of New York State. Some time ago I met a lovely lady, Holly, at a poetry reading which I attended and participated in. I am quite enamored of her and she seems to return my feelings. How the relationship will develop remains to be seen.

Not too long ago, I met Sylvia Bullett, a tall and strikingly interesting-looking lady at a concert, in which we both participated. She is a pianist and singer. She came to my 95th birthday party, hosted by my dear friends Sylvie and Wayne. She wrote me the following birthday poem, which strikes me as extremely creative (note that I was born in 1919.)

Smullyan, Smullyan, riddle me this:
What is the Mobius without the twist?
Without something can nothing exist?
Can a year be measured by a trip around the sun?
Could it be we were zero before we were one?
Was it ever thus and do numbers ever lie?
Can our souls be infinite like the decimal expansion of pi?
And if I were to count your years
— each one equaling five — I'd say
"Congratulations! for these 19 years alive!
Happy Birthday!"

* * *

I spend my days writing math and other books, reading, investigating this or that and communicating with others on the Internet, and making videos which combine my piano playing with my photography. I also go about town entertaining people in my irrepressible manner. You see, I have become so habituated in my early days of professional magic, that when I see a friendly group of people at a restaurant, I cannot resist going over and entertaining them. I have formed many friendships as a result.

Printed in the United States
By Bookmasters